AN INTRODUCTION TO INFORMATION THEORY

Symbols, Signals & Noise

JOHN R. PIERCE
Professor of Engineering
California Institute of Technology

Second, Revised Edition

Dover Publications, Inc.
New York

TO CLAUDE AND BETTY SHANNON

This Dover edition, first published in 1980, is an unabridged and revised version of the work originally published in 1961 by Harper & Brothers under the title *Symbols, Signals and Noise: The Nature and Process of Communication.*

International Standard Book Number: 0-486-24061-4
Library of Congress Catalog Card Number: 80-66678

Manufactured in the United States of America
Dover Publications, Inc.
31 East 2nd Street, Mineola, N.Y. 11501

Contents

Preface to the Dover Edition

THE REPUBLICATION OF THIS BOOK gave me an opportunity to correct and bring up to date *Symbols, Signals and Noise,*[1] which I wrote almost twenty years ago. Because the book deals largely with Shannon's work, which remains eternally valid, I found that there were not many changes to be made. In a few places I altered tense in referring to men who have died. I did not try to replace *cycles per second* (*cps*) by the more modern term, *hertz* (*hz*) nor did I change everywhere *communication theory* (Shannon's term) to *information theory*, the term I would use today.

Some things I did alter, rewriting a few paragraphs and about twenty pages without changing the pagination.

In Chapter X, Information Theory and Physics, I replaced a background radiation temperature of space of "2° to 4°K" (Heaven knows where I got that) by the correct value of 3.5°K, as determined by Penzias and Wilson. To the fact that in the absence of noise we can in principle transmit an unlimited number of bits per quantum, I added new material on quantum effects in communication.[2] I also replaced an obsolete thought-up example of space communication by a brief analysis of the microwave transmission of picture signals from the Voyager near Jupiter, and by an exposition of new possibilities.

[1] Harper Modern Science Series, Harper and Brothers, New York, 1961.

[2] See *Introduction to Communication Science and Systems*, John R. Pierce and Edward C. Posner, Plenum Publishing Corporation, New York, 1980.

In Chapter VII, Efficient Encoding, I rewrote a few pages concerning efficient source encoding of TV and changed a few sentences about pulse code modulation and about vocoders. I also changed the material on error correcting codes.

In Chapter XI, Cybernetics, I rewrote four pages on computers and programming, which have advanced incredibly during the last twenty years.

Finally, I made a few changes in the last short Chapter XIV, Back to Communication Theory.

Beyond these revisions, I call to the reader's attention a series of papers on the history of information theory that were published in 1973 in the *IEEE Transactions on Information Theory*[3] and two up-to-date books as telling in more detail the present state of information theory and the mathematical aspects of communication.[2,4,5]

Several chapters in the original book deal with areas relevant only through application or attempted application of information theory.

I think that Chapter XII, Information Theory and Psychology, gives a fair idea of the sort of applications attempted in that area. Today psychologists are less concerned with information theory than with cognitive science, a heady association of truly startling progress in the understanding of the nervous system, with ideas drawn from anthropology, linguistics and a belief that some powerful and simple mathematical order must underly human function. Cognitive science of today reminds me of cybernetics of twenty years ago.

As to Information Theory and Art, today the computer has replaced information theory in casual discussions. But, the ideas explored in Chapter XIII have been pursued further. I will mention some attractive poems produced by Marie Borroff[6,7], and, es-

[3] *IEEE Transactions on Information Theory*, Vol. IT–19, pp. 3–8, 145–148, 257–262, 381–389 (1973).

[4] *The Theory of Information and Coding*, Robert J. McEliece, Addison-Wesley, Reading, MA, 1977.

[5] *Principles of Digital Communication and Coding*, Andrew J. Viterbi and Jim K. Omura, McGraw Hill, New York, 1979.

[6] "Computer as Poet," Marie Borroff, *Yale Alumni Magazine*, Jan. 1971.

[7] *Computer Poems*, gathered by Richard Bailey, Potagannissing Press, 1973.

pecially a grammar of Swedish folksongs by means of which Johan Sundberg produced a number of authentic sounding tunes.[8]

This brings us back to language and Chapter VI, Language and Meaning. The problems raised in that chapter have not been resolved during the last twenty years. We do not have a complete grammar of any natural language. Indeed, formal grammar has proved most powerful in the area of computer languages. It is my reading that attention in linguistics has shifted somewhat to the phonological aspects of spoken language, to understanding what its building blocks are and how they interact—matters of great interest in the computer generation of speech from text. Chomsky and Halle have written a large book on stress,[9] and Liberman and Prince a smaller and very powerful account.[10]

So much for changes from the original *Signals, Symbols and Noise*. Beyond this, I can only reiterate some of the things I said in the preface to that book.

When James R. Newman suggested to me that I write a book about communication I was delighted. All my technical work has been inspired by one aspect or another of communication. Of course I would like to tell others what seems to me to be interesting and challenging in this important field.

It would have been difficult to do this and to give any sense of unity to the account before 1948 when Claude E. Shannon published "A Mathematical Theory of Communication."[11] Shannon's communication theory, which is also called information theory, has brought into a reasonable relation the many problems that have been troubling communication engineers for years. It has created a broad but clearly defined and limited field where before there were many special problems and ideas whose interrelations were not well

[8] "Generative Theories in Language and Musical Descriptions," Johan Sundberg and Bjorn Lindblom, *Cognition*, Vol. 4, pp. 99–122, 1976.

[9] *The Sound Pattern of English*, N. Chomsky and M. Halle, Harper and Row, 1968.

[10] "On Stress and Linguistic Rhythm," Mark Liberman and Alan Prince, *Linguistic Inquiry*, Vol. 8, No. 2, pp. 249–336, Spring, 1977.

[11] The papers, originally published in the *Bell System Technical Journal*, are reprinted in *The Mathematical Theory of Communication*, Shannon and Weaver, University of Illinois Press, first printing 1949. Shannon presented a somewhat different approach (used in Chapter IX of this book) in "Communication in the Presence of Noise," *Proceedings of the Institute of Radio Engineers*, Vol. 37, pp. 10–21, 1949.

understood. No one can accuse me of being a Shannon worshiper and get away unrewarded.

Thus, I felt that my account of communication must be an account of information theory as Shannon formulated it. The account would have to be broader than Shannon's in that it would discuss the relation, or lack of relation, of information theory to the many fields to which people have applied it. The account would have to be broader than Shannon's in that it would have to be less mathematical.

Here came the rub. My account could be *less* mathematical than Shannon's, but it could not be *nonmathematical.* Information theory is a mathematical theory. It starts from certain premises that define the aspects of communication with which it will deal, and it proceeds from these premises to various logical conclusions. The glory of information theory lies in certain mathematical theorems which are both surprising and important. To talk about information theory without communicating its real mathematical content would be like endlessly telling a man about a wonderful composer yet never letting him hear an example of the composer's music.

How was I to proceed? It seemed to me that I had to make the book self-contained, so that any mathematics in it could be understood without referring to other books or without calling for the particular content of early mathematical training, such as high school algebra. Did this mean that I had to avoid mathematical notation? Not necessarily, but any mathematical notation would have to be explained in the most elementary terms. I have done this both in the text and in an appendix; by going back and forth between the two, the mathematically untutored reader should be able to resolve any difficulties.

But just how difficult should the most difficult mathematical arguments be? Although it meant sliding over some very important points, I resolved to keep things easy compared with, say, the more difficult parts of Newman's *The World of Mathematics.* When the going is very difficult, I have merely indicated the general nature of the sort of mathematics used rather than trying to describe its content clearly.

Nonetheless, this book has sections which will be hard for the nonmathematical reader. I advise him merely to skim through these, gathering what he can. When he has gone through the book in this manner, he will see why the difficult sections are there. Then he can turn back and restudy them if he wishes. But, had I not put these difficult sections in, and had the reader wanted the sort of understanding that takes real thought, he would have been stuck. As far as I know, other available literature on information theory is either too simple or too difficult to help the diligent but inexpert reader beyond the easier parts of this book. I might note also that some of the literature is confused and some of it is just plain wrong.

By this sort of talk I may have raised wonder in the reader's mind as to whether or not information theory is really worth so much trouble, either on his part, for that matter, or on mine. I can only say that to the degree that the whole world of science and technology around us is important, information theory is important, for it is an important part of that world. To the degree to which an intelligent reader wants to know something both about that world and about information theory, it is worth his while to try to get a clear picture. Such a picture must show information theory neither as something utterly alien and unintelligible nor as something that can be epitomized in a few easy words and appreciated without effort.

The process of writing this book was not easy. Of course it could never have been written at all but for the work of Claude Shannon, who, besides inspiring the book through his work, read the original manuscript and suggested several valuable changes. David Slepian jolted me out of the rut of error and confusion in an even more vigorous way. E. N. Gilbert deflected me from error in several instances. Milton Babbitt reassured me concerning the major contents of the chapter on information theory and art and suggested a few changes. P. D. Bricker, H. M. Jenkins, and R. N. Shepard advised me in the field of psychology, but the views I finally expressed should not be attributed to them. The help of M. V. Mathews was invaluable. Benoit Mandelbrot helped me with Chapter XII. J. P. Runyon read the manuscript with care, and Eric Wolman uncovered an appalling number of textual errors, and made valuable suggestions as well. I am also indebted to Prof. Martin Harwit, who persuaded me and Dover that the book was worth reissuing. The reader is indebted to James R. Newman for the fact that I have provided a glossary, summaries at the ends of some chapters, and for my final attempts to make some difficult points a little clearer. To all of these I am indebted and not less to Miss F. M. Costello, who triumphed over the chaos of preparing and correcting the manuscript and figures. In preparing this new edition, I owe much to my secretary, Mrs. Patricia J. Neill.

September, 1979 J. R. PIERCE

CHAPTER I *The World and Theories*

IN 1948, CLAUDE E. SHANNON published a paper called "A Mathematical Theory of Communication"; it appeared in book form in 1949. Before that time, a few isolated workers had from time to time taken steps toward a general theory of communication. Now, thirty years later, communication theory, or information theory as it is sometimes called, is an accepted field of research. Many books on communication theory have been published, and many international symposia and conferences have been held. The Institute of Electrical and Electronic Engineers has a professional group on information theory, whose *Transactions* appear six times a year. Many other journals publish papers on information theory.

All of us use the words communication and information, and we are unlikely to underestimate their importance. A modern philosopher, A. J. Ayer, has commented on the wide meaning and importance of communication in our lives. We communicate, he observes, not only information, but also knowledge, error, opinions, ideas, experiences, wishes, orders, emotions, feelings, moods. Heat and motion can be communicated. So can strength and weakness and disease. He cites other examples and comments on the manifold manifestations and puzzling features of communication in man's world.

Surely, communication being so various and so important, a

theory of communication, a theory of generally accepted soundness and usefulness, must be of incomparable importance to all of us. When we add to *theory* the word *mathematical,* with all its implications of rigor and magic, the attraction becomes almost irresistible. Perhaps if we learn a few formulae our problems of communication will be solved, and we shall become the masters of information rather than the slaves of misinformation.

Unhappily, this is not the course of science. Some 2,300 years ago, another philosopher, Aristotle, discussed in his *Physics* a notion as universal as that of communication, that is, motion.

Aristotle defined motion as the fulfillment, insofar as it exists potentially, of that which exists potentially. He included in the concept of motion the increase and decrease of that which can be increased or decreased, coming to and passing away, and also being built. He spoke of three categories of motion, with respect to magnitude, affection, and place. He found, indeed, as he said, as many types of motion as there are meanings of the word *is.*

Here we see motion in all its manifest complexity. The complexity is perhaps a little bewildering to us, for the associations of words differ in different languages, and we would not necessarily associate motion with all the changes of which Aristotle speaks.

How puzzling this universal matter of motion must have been to the followers of Aristotle. It remained puzzling for over two millennia, until Newton enunciated the laws which engineers still use in designing machines and astronomers in studying the motions of stars, planets, and satellites. While later physicists have found that Newton's laws are only the special forms which more general laws assume when velocities are small compared with that of light and when the scale of the phenomena is large compared with the atom, they are a living part of our physics rather than a historical monument. Surely, when motion is so important a part of our world, we should study Newton's laws of motion. They say:

1. A body continues at rest or in motion with a constant velocity in a straight line unless acted upon by a force.

2. The change in velocity of a body is in the direction of the force acting on it, and the magnitude of the change is proportional to the force acting on the body times the time during which the force acts, and is inversely proportional to the mass of the body.

3. Whenever a first body exerts a force on a second body, the second body exerts an equal and oppositely directed force on the first body.

To these laws Newton added the universal law of gravitation:

4. Two particles of matter attract one another with a force acting along the line connecting them, a force which is proportional to the product of the masses of the particles and inversely proportional to the square of the distance separating them.

Newton's laws brought about a scientific and a philosophical revolution. Using them, Laplace reduced the solar system to an explicable machine. They have formed the basis of aviation and rocketry, as well as of astronomy. Yet, they do little to answer many of the questions about motion which Aristotle considered. Newton's laws solved the problem of motion as Newton defined it, not of motion in all the senses in which the word could be used in the Greek of the fourth century before our Lord or in the English of the twentieth century after.

Our speech is adapted to our daily needs or, perhaps, to the needs of our ancestors. We cannot have a separate word for every distinct object and for every distinct event; if we did we should be forever coining words, and communication would be impossible. In order to have language at all, many things or many events must be referred to by one word. It is natural to say that both men and horses run (though we may prefer to say that horses gallop) and convenient to say that a motor runs and to speak of a run in a stocking or a run on a bank.

The unity among these concepts lies far more in our human language than in any physical similarity with which we can expect science to deal easily and exactly. It would be foolish to seek some elegant, simple, and useful scientific theory of running which would embrace runs of salmon and runs in hose. It would be equally foolish to try to embrace in one theory all the motions discussed by Aristotle or all the sorts of communication and information which later philosophers have discovered.

In our everyday language, we use words in a way which is convenient in our everyday business. Except in the study of language itself, science does not seek understanding by studying words and their relations. Rather, science looks for things in nature, including our human nature and activities, which can be grouped together and understood. Such understanding is an ability to see what complicated or diverse events really do have in common (the planets in the heavens and the motions of a whirling skater on ice, for instance) and to describe the behavior accurately and simply.

The words used in such scientific descriptions are often drawn from our everyday vocabulary. Newton used force, mass, velocity, and attraction. When used in science, however, a particular meaning is given to such words, a meaning narrow and often new. We cannot discuss in Newton's terms force of circumstance, mass media, or the attraction of Brigitte Bardot. Neither should we expect that communication theory will have something sensible to say about every question we can phrase using the words communication or information.

A valid scientific theory seldom if ever offers the solution to the pressing problems which we repeatedly state. It seldom supplies a sensible answer to our multitudinous questions. Rather than rationalizing our ideas, it discards them entirely, or, rather, it leaves them as they were. It tells us in a fresh and new way what aspects of our experience can profitably be related and simply understood. In this book, it will be our endeavor to seek out the ideas concerning communication which can be so related and understood.

When the portions of our experience which can be related have been singled out, and when they have been related and understood, we have a *theory* concerning these matters. Newton's laws of motion form an important part of *theoretical physics,* a field called *mechanics.* The laws themselves are not the whole of the theory; they are merely the basis of it, as the axioms or postulates of geometry are the basis of geometry. The theory embraces both the assumptions themselves and the mathematical working out of the logical consequences which must necessarily follow from the assumptions. Of course, these consequences must be in accord with the complex phenomena of the world about us if the theory is to be a valid theory, and an invalid theory is useless.

The ideas and assumptions of a theory determine the *generality* of the theory, that is, to how wide a range of phenomena the theory applies. Thus, Newton's laws of motion and of gravitation are very general; they explain the motion of the planets, the timekeeping properties of a pendulum, and the behavior of all sorts of machines and mechanisms. They do not, however, explain radio waves.

Maxwell's equations[1] explain all (non-quantum) electrical phenomena; they are very general. A branch of electrical theory called *network theory* deals with the electrical properties of electrical circuits, or networks, made by interconnecting three sorts of idealized electrical structures: resistors (devices such as coils of thin, poorly conducting wire or films of metal or carbon, which impede the flow of current), inductors (coils of copper wire, sometimes wound on magnetic cores), and capacitors (thin sheets of metal separated by an insulator or dielectric such as mica or plastic; the Leyden jar was an early form of capacitor). Because network theory deals only with the electrical behavior of certain specialized and idealized physical structures, while Maxwell's equations describe the electrical behavior of any physical structure, a physicist would say that network theory is *less* general than are Maxwell's equations, for Maxwell's equations cover the behavior not only of idealized electrical networks but of all physical structures and include the behavior of radio waves, which lies outside of the scope of network theory.

Certainly, the most general theory, which explains the greatest range of phenomena, is the most powerful and the best; it can always be specialized to deal with simple cases. That is why physicists have sought a unified field theory to embrace mechanical laws and gravitation and all electrical phenomena. It might, indeed, seem that all theories could be ranked in order of generality, and, if this is possible, we should certainly like to know the place of communication theory in such a hierarchy.

Unfortunately, life isn't as simple as this. In one sense, network theory is less general than Maxwell's equations. In another sense,

[1] In 1873, in his treatise *Electrictity and Magnetism,* James Clerk Maxwell presented and fully explained for the first time the natural laws relating electric and magnetic fields and electric currents. He showed that there should be *electromagnetic waves* (radio waves) which travel with the speed of light. Hertz later demonstrated these experimentally, and we now know that light is electromagnetic waves. Maxwell's equations are the mathematical statement of Maxwell's theory of electricity and magnetism. They are the foundation of all electric art.

however, it is more general, for all the mathematical results of network theory hold for vibrating mechanical systems made up of idealized mechanical components as well as for the behavior of interconnections of idealized electrical components. In mechanical applications, a spring corresponds to a capacitor, a mass to an inductor, and a dashpot or damper, such as that used in a door closer to keep the door from slamming, corresponds to a resistor. In fact, network theory might have been developed to explain the behavior of mechanical systems, and it is so used in the field of acoustics. The fact that network theory evolved from the study of idealized electrical systems rather than from the study of idealized mechanical systems is a matter of history, not of necessity.

Because all of the mathematical results of network theory apply to certain specialized and idealized mechanical systems, as well as to certain specialized and idealized electrical systems, we can say that in a sense network theory is *more* general than Maxwell's equations, which do not apply to mechanical systems at all. In another sense, of course, Maxwell's equations are more general than network theory, for Maxwell's equations apply to all electrical systems, not merely to a specialized and idealized class of electrical circuits.

To some degree we must simply admit that this is so, without being able to explain the fact fully. Yet, we can say this much. Some theories are very strongly *physical* theories. Newton's laws and Maxwell's equations are such theories. Newton's laws deal with mechanical phenomena; Maxwell's equations deal with electrical phenomena. Network theory is essentially a *mathematical* theory. The terms used in it can be given various physical meanings. The theory has interesting things to say about different physical phenomena, about mechanical as well as electrical vibrations.

Often a mathematical theory is the offshoot of a physical theory or of physical theories. It can be an elegant mathematical formulation and treatment of certain aspects of a general physical theory. Network theory is such a treatment of certain physical behavior common to electrical and mechanical devices. A branch of mathematics called *potential theory* treats problems common to electric, magnetic, and gravitational fields and, indeed, in a degree to aerodynamics. Some theories seem, however, to be more mathematical than physical in their very inception.

We use many such mathematical theories in dealing with the physical world. Arithmetic is one of these. If we label one of a group of apples, dogs, or men 1, another 2, and so on, and if we have used up just the first 16 numbers when we have labeled all members of the group, we feel confident that the group of objects can be divided into two equal groups each containing 8 objects ($16 \div 2 = 8$) or that the objects can be arranged in a square array of four parallel rows of four objects each (because 16 is a perfect square; $16 = 4 \times 4$). Further, if we line the apples, dogs, or men up in a row, there are 2,092,278,988,800 possible sequences in which they can be arranged, corresponding to the 2,092,278,-988,800 different sequences of the integers 1 through 16. If we used up 13 rather than 16 numbers in labeling the complete collection of objects, we feel equally certain that the collection could not be divided into any number of equal heaps, because 13 is a prime number and cannot be expressed as a product of factors.

This seems not to depend at all on the nature of the objects. Insofar as we can assign numbers to the members of any collection of objects, the results we get by adding, subtracting, multiplying, and dividing numbers or by arranging the numbers in sequence hold true. The connection between numbers and collections of objects seems so natural to us that we may overlook the fact that arithmetic is itself a mathematical theory which can be applied to nature only to the degree that the properties of numbers correspond to properties of the physical world.

Physicists tell us that we can talk sense about the total number of a group of elementary particles, such as electrons, but we can't assign particular numbers to particular particles because the particles are in a very real sense indistinguishable. Thus, we can't talk about arranging such particles in different orders, as numbers can be arranged in different sequences. This has important consequences in a part of physics called *statistical mechanics.* We may also note that while Euclidean geometry is a mathematical theory which serves surveyors and navigators admirably in their practical concerns, there is reason to believe that Euclidean geometry is not quite accurate in describing astronomical phenomena.

How can we describe or classify theories? We can say that a theory is very narrow or very general in its scope. We can also distinguish theories as to whether they are strongly physical or strongly mathematical. Theories are strongly physical when they describe very completely some range of physical phenomena, which in practice is always limited. Theories become more mathematical or abstract when they deal with an idealized class of phenomena or with only certain aspects of phenomena. Newton's laws are strongly physical in that they afford a complete description of mechanical phenomena such as the motions of the planets or the behavior of a pendulum. Network theory is more toward the mathematical or abstract side in that it is useful in dealing with a variety of idealized physical phenomena. Arithmetic is very mathematical and abstract; it is equally at home with one particular property of many sorts of physical entities, with numbers of dogs, numbers of men, and (if we remember that electrons are indistinguishable) with numbers of electrons. It is even useful in reckoning numbers of days.

In these terms, communication theory is both very strongly mathematical and quite general. Although communication theory grew out of the study of electrical communication, it attacks problems in a very abstract and general way. It provides, in the *bit,* a universal measure of amount of information in terms of choice or uncertainty. Specifying or learning the choice between two equally probable alternatives, which might be messages or numbers to be transmitted, involves one bit of information. Communication theory tells us how many bits of information can be sent per second over perfect and imperfect communication channels in terms of rather abstract descriptions of the properties of these channels. Communication theory tells us how to measure the rate at which a message source, such as a speaker or a writer, generates information. Communication theory tells us how to represent, or *encode,* messages from a particular message source efficiently for transmission over a particular sort of channel, such as an electrical circuit, and it tells us when we can avoid errors in transmission.

Because communication theory discusses such matters in very general and abstract terms, it is sometimes difficult to use the understanding it gives us in connection with particular, practical problems. However, because communication theory has such an abstract and general mathematical form, it has a very broad field of application. Communication theory is useful in connection with written and spoken language, the electrical and mechanical transmission of messages, the behavior of machines, and, perhaps, the behavior of people. Some feel that it has great relevance and importance to physics in a way that we shall discuss much later in this book.

Primarily, however, communication theory is, as Shannon described it, a *mathematical* theory of communication. The concepts are formulated in mathematical terms, of which widely different physical examples can be given. Engineers, psychologists, and physicists may use communication theory, but it remains a mathematical theory rather than a physical or psychological theory or an engineering art.

It is not easy to present a mathematical theory to a general audience, yet communication theory is a mathematical theory, and to pretend that one can discuss it while avoiding mathematics entirely would be ridiculous. Indeed, the reader may be startled to find equations and formulae in these pages; these state accurately ideas which are also described in words, and I have included an appendix on mathematical notation to help the nonmathematical reader who wants to read the equations aright.

I am aware, however, that mathematics calls up chiefly unpleasant pictures of multiplication, division, and perhaps square roots, as well as the possibly traumatic experiences of high-school classrooms. This view of mathematics is very misleading, for it places emphasis on special notation and on tricks of manipulation, rather than on the aspect of mathematics that is most important to mathematicians. Perhaps the reader has encountered theorems and proofs in geometry; perhaps he has not encountered them at all, yet theorems and proofs are of primary importance in all mathematics, pure and applied. The important results of information theory are stated in the form of mathematical theorems, and these *are* theorems only because it is possible to prove that they are true statements.

Mathematicians start out with certain assumptions and definitions, and then by means of mathematical arguments or proofs they are able to show that certain statements or theorems are true. This is what Shannon accomplished in his "Mathematical Theory of Communication." The truth of a theorem depends on the validity of the assumptions made and on the validity of the argument or proof which is used to establish it.

All of this is pretty abstract. The best way to give some idea of the meaning of *theorem* and *proof* is certainly by means of examples. I cannot do this by asking the general reader to grapple, one by one and in all their gory detail, with the difficult theorems of communication theory. Really to understand thoroughly the proofs of such theorems takes time and concentration even for one with some mathematical background. At best, we can try to get at the content, meaning, and importance of the theorems.

The expedient I propose to resort to is to give some examples of simpler mathematical theorems and their proof. The first example concerns a game called *hex,* or *Nash.* The theorem which will be proved is that the player with first move can win.

Hex is played on a board which is an array of forty-nine hexagonal cells or spaces, as shown in Figure I-1, into which markers may be put. One player uses black markers and tries to place them so as to form a continuous, if wandering, path between the black area at the left and the black area at the right. The other player uses white markers and tries to place them so as to form a continuous, if wandering, path between the white area at the top and the white area at the bottom. The players play alternately, each placing one marker per play. Of course, one player has to start first.

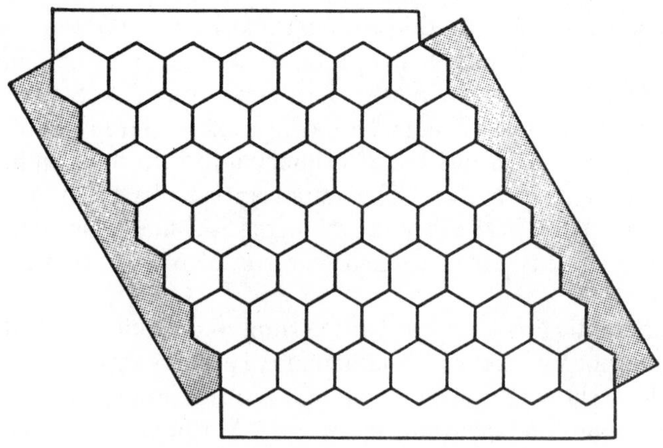

Fig. I-1

In order to prove that the first player can win, it is necessary first to prove that when the game is played out, so that there is either a black or a white marker in each cell, one of the players must have won.

Theorem I: *Either one player or the other wins.*

Discussion: In playing some games, such as chess and ticktacktoe, it may be that neither player will win, that is, that the game will end in a draw. In matching heads or tails, one or the other necessarily wins. What one must show to prove this theorem is that, when each cell of the hex board is covered by either a black or a white marker, either there must be a black path between the black areas which will interrupt any possible white path between the white areas or there must be a white path between the white areas which will interrupt any possible black path between the black areas, so that either white or black must have won.

Proof: Assume that each hexagon has been filled in with either a black or a white marker. Let us start from the left-hand corner of the upper white border, point I of Figure I-2, and trace out the boundary between white and black hexagons or borders. We will proceed always along a side with black on our right and white on our left. The boundary so traced out will turn at the successive corners, or vertices, at which the sides of hexagons meet. At a corner, or vertex, we can have only two essentially different con-

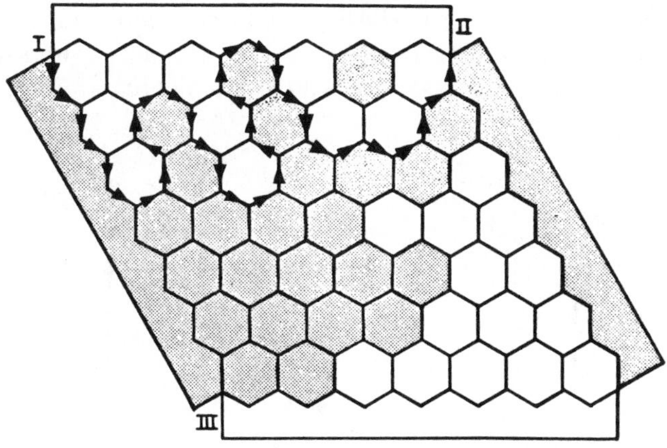

Fig. I-2

ditions. Either there will be two touching black hexagons on the right and one white hexagon on the left, as in *a* of Figure I-3, or two touching white hexagons on the left and one black hexagon on the right, as shown in *b* of Figure I-3. We note that in either case there will be a continuous black path to the right of the boundary and a continuous white path to the left of the boundary. We also note that in neither *a* nor *b* of Figure I-3 can the boundary cross or join itself, because only one path through the vertex has black on the right and white on the left. We can see that these two facts are true for boundaries between the black and white borders and hexagons as well as for boundaries between black and white hexagons. Thus, along the left side of the boundary there must be a continuous path of white hexagons to the upper white border, and along the right side of the boundary there must be a continuous path of black hexagons to the left black border. As the boundary cannot cross itself, it cannot circle indefinitely, but must eventually reach a black border or a white border. If the boundary reaches a black border or white border with black on its right and white on its left, as we have prescribed, at any place except corner II or corner III, we can extend the boundary further with black on its right and white on its left. Hence, the boundary will reach either point II or point III. If it reaches point II, as shown in Figure I-2, the black hexagons on the right, which are connected to the left black border, will also be connected to the right black border, while the white hexagons to the left will be connected to the upper white border only, and black will have won. It is clearly impossible for white to have won also, for the continuous band of adjacent

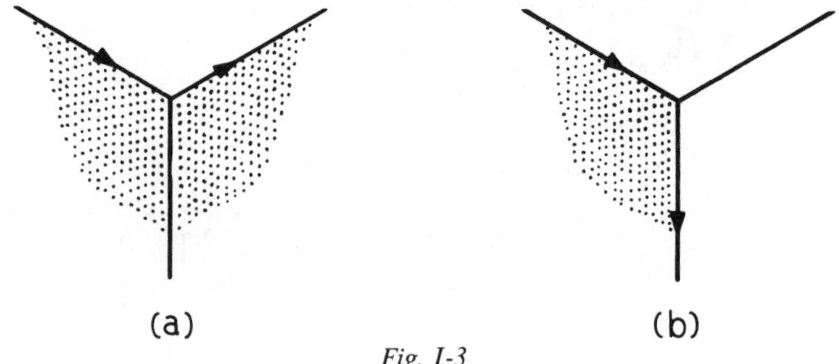

Fig. I-3

black cells from the left border to the right precludes a continuous band of white cells to the bottom border. We see by similar argument that, if the boundary reaches point III, white will have won.

Theorem II: *The player with the first move can win.*

Discussion: By *can* is meant that there exists a way, if only the player were wise enough to know it. The method for winning would consist of a particular first move (more than one might be allowable but are not necessary) and a chart, formula, or other specification or recipe giving a correct move following any possible move made by his opponent at any subsequent stage of the game, such that if, each time he plays, the first player makes the prescribed move, he will win regardless of what moves his opponent may make.

Proof: Either there must be some way of play which, if followed by the first player, will insure that he wins or else, no matter how the first player plays, the second player must be able to choose moves which will preclude the first player from winning, so that he, the second player, will win. Let us assume that the player with the second move does have a sure recipe for winning. Let the player with the first move make his first move in any way, and then, after his opponent has made one move, let the player with the first move apply the hypothetical recipe which is supposed to allow the player with the second move to win. If at any time a move calls for putting a piece on a hexagon occupied by a piece he has already played, let him place his piece instead on any unoccupied space. The designated space will thus be occupied. The fact that by starting first he has an extra piece on the board may keep his opponent from occupying a particular hexagon but not the player with the extra piece. Hence, the first player can occupy the hexagons designated by the recipe and must win. This is contrary to the original assumption that the player with the second move can win, and so this assumption must be false. Instead, it must be possible for the player with the first move to win.

A mathematical purist would scarcely regard these proofs as rigorous in the form given. The proof of theorem II has another curious feature; it is not a *constructive* proof. That is, it does not show the player with the first move, who can win in principle, how to go about winning. We will come to an example of a constructive proof in a moment. First, however, it may be appropriate to philosophize a little concerning the nature of theorems and the need for proving them.

Mathematical theorems are inherent in the rigorous statement of the general problem or field. That the player with the first move can win at hex is necessarily so once the game and its rules of play have been specified. The theorems of Euclidean geometry are necessarily so because of the stated postulates.

With sufficient intelligence and insight, we could presumably see the truth of theorems immediately. The young Newton is said to have found Euclid's theorems obvious and to have been impatient with their proofs.

Ordinarily, while mathematicians may suspect or conjecture the truth of certain statements, they have to prove theorems in order to be certain. Newton himself came to see the importance of proof, and he proved many new theorems by using the methods of Euclid.

By and large, mathematicians have to proceed step by step in attaining sure knowledge of a problem. They laboriously prove one theorem after another, rather than seeing through everything in a flash. Too, they need to prove the theorems in order to convince others.

Sometimes a mathematician needs to prove a theorem to convince himself, for the theorem may seem contrary to common sense. Let us take the following problem as an example: Consider the square, 1 inch on a side, at the left of Figure I-4. We can specify any point in the square by giving two numbers, y, the height of the point above the base of the square, and x, the distance of the point from the left-hand side of the square. Each of these numbers will be less than one. For instance, the point shown will be represented by

$x = 0.547000\ldots$ (ending in an endless sequence of zeros)
$y = 0.312000\ldots$ (ending in an endless sequence of zeros)

Suppose we pair up points on the square with points on the line, so that every point on the line is paired with just one point on the square and every point on the square with just one point on the line. If we do this, we are said to have *mapped* the square onto the line in a *one-to-one* way, or to have achieved a *one-to-one mapping* of the square onto the line.

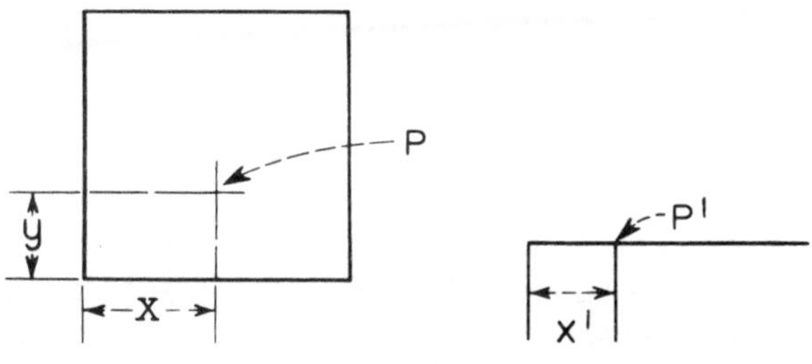

Fig. I-4

Theorem: *It is possible to map a square of unit area onto a line of unit length in a one-to-one way.*[2]

Proof: Take the successive digits of the height of the point in the square and let them form the first, third, fifth, and so on digits of a number x'. Take the digits of the distance of the point P from the left side of the square, and let these be the second, fourth, sixth, etc., of the digits of the number x'. Let x' be the distance of the point P' from the left-hand end of the line. Then the point P' maps the point P of the square onto the line uniquely, in a one-to-one way. We see that changing either x or y will change x' to a new and appropriate number, and changing x' will change x and y. To each point x,y in the square corresponds just one point x' on the line, and to each point x' on the line corresponds just one point x,y in the square, the requirement for one-to-one mapping.[3]

In the case of the example given before

$$x = 0.547000\ldots$$
$$y = 0.312000\ldots$$
$$x' = 0.351427000\ldots$$

In the case of most points, including those specified by irrational numbers, the endless string of digits representing the point will not become a sequence of zeros nor will it ever repeat.

Here we have an example of a constructive proof. We show that we can map each point of a square into a point on a line segment in a one-to-one way by giving an explicit recipe for doing this. Many mathematicians prefer constructive proofs to proofs which

[2] This has been restricted for convenience; the size doesn't matter.

[3] This proof runs into resolvable difficulties in the case of some numbers such as ½, which can be represented decimally .5 followed by an infinite sequence of zeros or .4 followed by an infinite sequence of nines.

are not constructive, and mathematicians of the intuitionist school reject nonconstructive proofs in dealing with infinite sets, in which it is impossible to examine all the members individually for the property in question.

Let us now consider another matter concerning the mapping of the points of a square on a line segment. Imagine that we move a pointer along the line, and imagine a pointer simultaneously moving over the face of the square so as to point out the points in the square corresponding to the points that the first pointer indicates on the line. We might imagine (contrary to what we shall prove) the following: If we moved the first pointer slowly and smoothly along the line, the second pointer would move slowly and smoothly over the face of the square. All the points lying in a small cluster on the line would be represented by points lying in a small cluster on the face of the square. If we moved the pointer a short distance along the line, the other pointer would move a short distance over the face of the square, and if we moved the pointer a shorter distance along the line, the other pointer would move a shorter distance across the face of the square, and so on. If this were true we could say that the one-to-one mapping of the points of the square into points on the line was *continuous.*

However, it turns out that a one-to-one mapping of the points in a square into the points on a line cannot be continuous. As we move smoothly along a curve through the square, the points on the line which represent the successive points on the square *necessarily* jump around erratically, not only for the mapping described above but for any one-to-one mapping whatever. Any one-to-one mapping of the square onto the line is *discontinuous.*

Theorem: *Any one-to-one mapping of a square onto a line must be discontinuous.*

Proof: Assume that the one-to-one mapping is continuous. If this is to be so then all the points along some arbitrary curve AB of Figure I-5 on the square must map into the points lying between the corresponding points A' and B'. If they did not, in moving along the curve in the square we would either jump from one end of the line to the other (discontinuous mapping) or pass through one point on the line twice (not one-to-one mapping). Let us now choose a point C' to the left of line segment $A'B'$ and D' to the right of $A'B'$ and locate the corresponding points C and D in the

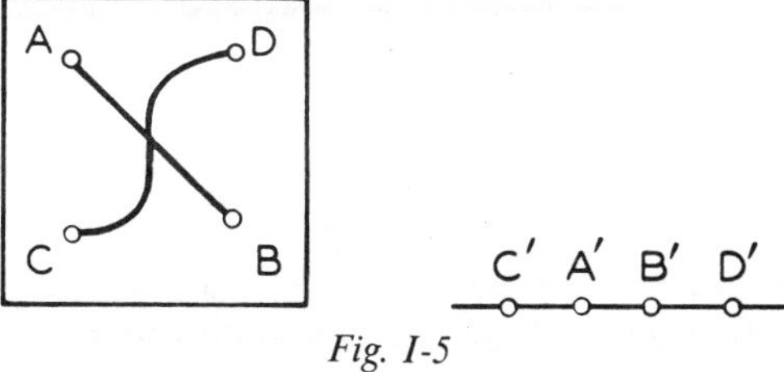

Fig. I-5

square. Draw a curve connecting C and D and crossing the curve from A to B. Where the curve crosses the curve AB it will have a point in common with AB; hence, this one point of CD must map into a point lying between A' and B', and all other points which are not on AB must map to points lying outside of $A'B'$, either to the left or the right of $A'B'$. This is contrary to our assumption that the mapping was continuous, and so the mapping cannot be continuous.

We shall find that these theorems, that the points of a square can be mapped onto a line and that the mapping is necessarily discontinuous, are both important in communication theory, so we have proved one theorem which, unlike those concerning hex, will be of some use to us.

Mathematics is a way of finding out, step by step, facts which are inherent in the statement of the problem but which are not immediately obvious. Usually, in applying mathematics one must first hit on the facts and then verify them by proof. Here we come upon a knotty problem, for the proofs which satisfied mathematicians of an earlier day do not satisfy modern mathematicians.

In our own day, an irascible minor mathematician who reviewed Shannon's original paper on communication theory expressed doubts as to whether or not the author's mathematical intentions were honorable. Shannon's theorems are true, however, and proofs have been given which satisfy even rigor-crazed mathematicians. The simple proofs which I have given above as illustrations of mathematics are open to criticism by purists.

What I have tried to do is to indicate the nature of mathematical reasoning, to give some idea of what a theorem is and of how it may be proved. With this in mind, we will go on to the mathematical theory of communication, its theorems, which we shall not really prove, and to some implications and associations which extend beyond anything that we can establish with mathematical certainty.

As I have indicated earlier in this chapter, communication theory as Shannon has given it to us deals in a very broad and abstract way with certain important problems of communication and information, but it cannot be applied to all problems which we can phrase using the words *communication* and *information* in their many popular senses. Communication theory deals with certain aspects of communication which *can* be associated and organized in a useful and fruitful way, just as Newton's laws of motion deal with mechanical motion only, rather than with all the named and indeed different phenomena which Aristotle had in mind when he used the word *motion*.

To succeed, science must attempt the possible. We have no reason to believe that we can unify all the things and concepts for which we use a common word. Rather we must seek that part of experience which can be related. When we have succeeded in relating certain aspects of experience we have a theory. Newton's laws of motion are a theory which we can use in dealing with mechanical phenomena. Maxwell's equations are a theory which we can use in connection with electrical phenomena. Network theory we can use in connection with certain simple sorts of electrical *or* mechanical devices. We can use arithmetic very generally in connection with numbers of men, stones, or stars, and geometry in measuring land, sea, or galaxies.

Unlike Newton's laws of motion and Maxwell's equations, which are strongly physical in that they deal with certain classes of physical phenomena, communication theory is abstract in that it applies to many sorts of communication, written, acoustical, or electrical. Communication theory deals with certain important but abstract aspects of communication. Communication theory proceeds from clear and definite assumptions to theorems concerning information sources and communication channels. In this it is essentially mathematical, and in order to understand it we must understand the idea of a theorem as a statement which must be proved, that is, which must be shown to be the necessary consequence of a set of initial assumptions. This is an idea which is the very heart of mathematics as mathematicians understand it.

CHAPTER II *The Origins of Information Theory*

MEN HAVE BEEN at odds concerning the value of history. Some have studied earlier times in order to find a universal system of the world, in whose inevitable unfolding we can see the future as well as the past. Others have sought in the past prescriptions for success in the present. Thus, some believe that by studying scientific discovery in another day we can learn how to make discoveries. On the other hand, one sage observed that we learn nothing from history except that we never learn anything from history, and Henry Ford asserted that history is bunk.

All of this is as far beyond me as it is beyond the scope of this book. I will, however, maintain that we can learn at least two things from the history of science.

One of these is that many of the most general and powerful discoveries of science have arisen, not through the study of phenomena as they occur in nature, but, rather, through the study of phenomena in man-made devices, in products of technology, if you will. This is because the phenomena in man's machines are simplified and ordered in comparison with those occurring naturally, and it is these simplified phenomena that man understands most easily.

Thus, the existence of the steam engine, in which phenomena involving heat, pressure, vaporization, and condensation occur in a simple and orderly fashion, gave tremendous impetus to the very powerful and general science of thermodynamics. We see this especially in the work of Carnot.[1] Our knowledge of aerodynamics and hydrodynamics exists chiefly because airplanes and ships exist, not because of the existence of birds and fishes. Our knowledge of electricity came mainly not from the study of lightning, but from the study of man's artifacts.

Similarly, we shall find the roots of Shannon's broad and elegant theory of communication in the simplified and seemingly easily intelligible phenomena of telegraphy.

The second thing that history can teach us is with what difficulty understanding is won. Today, Newton's laws of motion seem simple and almost inevitable, yet there was a day when they were undreamed of, a day when brilliant men had the oddest notions about motion. Even discoverers themselves sometimes seem incredibly dense as well as inexplicably wonderful. One might expect of Maxwell's treatise on electricity and magnetism a bold and simple pronouncement concerning the great step he had taken. Instead, it is cluttered with all sorts of such lesser matters as once seemed important, so that a naïve reader might search long to find the novel step and to restate it in the simple manner familiar to us. It is true, however, that Maxwell stated his case clearly elsewhere.

Thus, a study of the origins of scientific ideas can help us to value understanding more highly for its having been so dearly won. We can often see men of an earlier day stumbling along the edge of discovery but unable to take the final step. Sometimes we are tempted to take it for them and to say, because they stated many of the required concepts in juxtaposition, that they must really have reached the general conclusion. This, alas, is the same trap into which many an ungrateful fellow falls in his own life. When someone actually solves a problem that he merely has had ideas about, he believes that he understood the matter all along.

Properly understood, then, the origins of an idea can help to show what its real content is; what the degree of understanding was before the idea came along and how unity and clarity have been attained. But to attain such understanding we must trace the actual course of discovery, not some course which we feel discovery

[1] N. L. S. Carnot (1796–1832) first proposed an ideal expansion of gas (the *Carnot cycle*) which will extract the maximum possible mechanical energy from the thermal energy of the steam.

should or could have taken, and we must see problems (if we can) as the men of the past saw them, not as we see them today.

In looking for the origin of communication theory one is apt to fall into an almost trackless morass. I would gladly avoid this entirely but cannot, for others continually urge their readers to enter it. I only hope that they will emerge unharmed with the help of the following grudgingly given guidance.

A particular quantity called *entropy* is used in thermodynamics and in statistical mechanics. A quantity called *entropy* is used in communication theory. After all, thermodynamics and statistical mechanics are older than communication theory. Further, in a paper published in 1929, L. Szilard, a physicist, used an idea of information in resolving a particular physical paradox. From these facts we might conclude that communication theory somehow grew out of statistical mechanics.

This easy but misleading idea has caused a great deal of confusion even among technical men. Actually, communication theory evolved from an effort to solve certain problems in the field of electrical communication. Its entropy was called entropy by mathematical analogy with the entropy of statistical mechanics. The chief relevance of this entropy is to problems quite different from those which statistical mechanics attacks.

In thermodynamics, the entropy of a body of gas depends on its temperature, volume, and mass—and on what gas it is—just as the energy of the body of gas does. If the gas is allowed to expand in a cylinder, pushing on a slowly moving piston, with no flow of heat to or from the gas, the gas will become cooler, losing some of its thermal energy. This energy appears as work done on the piston. The work may, for instance, lift a weight, which thus stores the energy lost by the gas.

This is a *reversible* process. By this we mean that if work is done in pushing the piston slowly back against the gas and so recompressing it to its original volume, the exact original energy, pressure, and temperature will be restored to the gas. In such a reversible process, the entropy of the gas remains constant, while its energy changes.

Thus, entropy is an indicator of reversibility; when there is no change of entropy, the process is reversible. In the example discussed above, energy can be transferred repeatedly back and forth between thermal energy of the compressed gas and mechanical energy of a lifted weight.

Most physical phenomena are not reversible. Irreversible phenomena always involve an increase of entropy.

Imagine, for instance, that a cylinder which allows no heat flow in or out is divided into two parts by a partition, and suppose that there is gas on one side of the partition and none on the other. Imagine that the partition suddenly vanishes, so that the gas expands and fills the whole container. In this case, the thermal energy remains the same, but the entropy increases.

Before the partition vanished we could have obtained mechanical energy from the gas by letting it flow into the empty part of the cylinder through a little engine. After the removal of the partition and the subsequent increase in entropy, we cannot do this. The entropy can increase while the energy remains constant in other similar circumstances. For instance, this happens when heat flows from a hot object to a cold object. Before the temperatures were equalized, mechanical work could have been done by making use of the temperature difference. After the temperature difference has disappeared, we can no longer use it in changing part of the thermal energy into mechanical energy.

Thus, an increase in entropy means a decrease in our ability to change thermal energy, the energy of heat, into mechanical energy. An increase of entropy means a decrease of available energy.

While thermodynamics gave us the concept of entropy, it does not give a detailed physical picture of entropy, in terms of positions and velocities of molecules, for instance. *Statistical mechanics* does give a detailed mechanical meaning to entropy in particular cases. In general, the meaning is that an increase in entropy means a decrease in order. But, when we ask what order means, we must in some way equate it with knowledge. Even a very complex arrangement of molecules can scarcely be disordered if we know the position and velocity of every one. Disorder in the sense in which it is used in statistical mechanics involves unpredictability based on a lack of knowledge of the positions and velocities of molecules. Ordinarily we lack such knowledge when the arrangement of positions and velocities is "complicated."

Let us return to the example discussed above in which all the molecules of a gas are initially on one side of a partition in a cylinder. If the molecules are all on one side of the partition, and if we know this, the entropy is less than if they are distributed on both sides of the partition. Certainly, we know more about the positions of the molecules when we know that they are all on one side of the partition than if we merely know that they are somewhere within the whole container. The more detailed our knowledge is concerning a physical system, the less uncertainty we have concerning it (concerning the location of the molecules, for instance) and the less the entropy is. Conversely, more uncertainty means more entropy.

Thus, in physics, entropy is associated with the possibility of converting thermal energy into mechanical energy. If the entropy does not change during a process, the process is reversible. If the entropy increases, the available energy decreases. Statistical mechanics interprets an increase of entropy as a decrease in order or, if we wish, as a decrease in our knowledge.

The applications and details of entropy in physics are of course much broader than the examples I have given can illustrate, but I believe that I have indicated its nature and something of its importance. Let us now consider the quite different purpose and use of the entropy of communication theory.

In communication theory we consider a message source, such as a writer or a speaker, which may produce on a given occasion any one of many possible messages. The amount of information conveyed by the message increases as the amount of uncertainty as to what message actually will be produced becomes greater. A message which is one out of ten possible messages conveys a smaller amount of information than a message which is one out of a million possible messages. The entropy of communication theory is a measure of this uncertainty and the uncertainty, or entropy, is taken as the measure of the amount of information conveyed by a message from a source. The more we know about what message the source will produce, the less uncertainty, the less the entropy, and the less the information.

We see that the ideas which gave rise to the entropy of physics and the entropy of communication theory are quite different. One can be fully useful without any reference at all to the other. Nonetheless, both the entropy of statistical mechanics and that of communication theory can be described in terms of uncertainty, in similar mathematical terms. Can some significant and useful relation be established between the two different entropies and, indeed, between physics and the mathematical theory of communication?

Several physicists and mathematicians have been anxious to show that communication theory and its entropy are extremely important in connection with statistical mechanics. This is still a confused and confusing matter. The confusion is sometimes aggravated when more than one meaning of *information* creeps into a discussion. Thus, *information* is sometimes associated with the idea of *knowledge* through its popular use rather than with *uncertainty* and the resolution of uncertainty, as it is in communication theory.

We will consider the relation between communication theory and physics in Chapter X, after arriving at some understanding of communication theory. Here I will merely say that the efforts to marry communication theory and physics have been more interesting than fruitful. Certainly, such attempts have not produced important new results or understanding, as communication theory has in its own right.

Communication theory has its origins in the study of electrical communication, not in statistical mechanics, and some of the ideas important to communication theory go back to the very origins of electrical communication.

During a transatlantic voyage in 1832, Samuel F. B. Morse set to work on the first widely successful form of electrical telegraph. As Morse first worked it out, his telegraph was much more complicated than the one we know. It actually drew short and long lines on a strip of paper, and sequences of these represented, not the letters of a word, but numbers assigned to words in a dictionary or code book which Morse completed in 1837. This is (as we shall see) an efficient form of coding, but it is clumsy.

While Morse was working with Alfred Vail, the old coding was given up, and what we now know as the Morse code had been devised by 1838. In this code, letters of the alphabet are represented by spaces, dots, and dashes. The space is the absence of an electric current, the dot is an electric current of short duration, and the dash is an electric current of longer duration.

Various combinations of dots and dashes were cleverly assigned to the letters of the alphabet. E, the letter occurring most frequently in English text, was represented by the shortest possible code symbol, a single dot, and, in general, short combinations of dots and dashes were used for frequently used letters and long combinations for rarely used letters. Strangely enough, the choice was not guided by tables of the relative frequencies of various letters in English text nor were letters in text counted to get such data. Relative frequencies of occurrence of various letters were estimated by counting the number of types in the various compartments of a printer's type box!

We can ask, would some other assignment of dots, dashes, and spaces to letters than that used by Morse enable us to send English text faster by telegraph? Our modern theory tells us that we could only gain about 15 per cent in speed. Morse was very successful indeed in achieving his end, and he had the end clearly in mind. The lesson provided by Morse's code is that it matters profoundly how one translates a message into electrical signals. This matter is at the very heart of communication theory.

In 1843, Congress passed a bill appropriating money for the construction of a telegraph circuit between Washington and Baltimore. Morse started to lay the wire underground, but ran into difficulties which later plagued submarine cables even more severely. He solved his immediate problem by stringing the wire on poles.

The difficulty which Morse encountered with his underground wire remained an important problem. Different circuits which conduct a steady electric current equally well are not necessarily equally suited to electrical communication. If one sends dots and dashes too fast over an underground or undersea circuit, they are run together at the receiving end. As indicated in Figure II-1, when we send a short burst of current which turns abruptly on and off, we receive at the far end of the circuit a longer, smoothed-out rise and fall of current. This longer flow of current may overlap the current of another symbol sent, for instance, as an absence of current. Thus, as shown in Figure II-2, when a clear and distinct

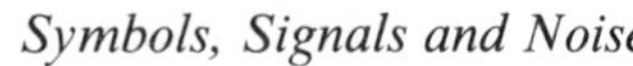

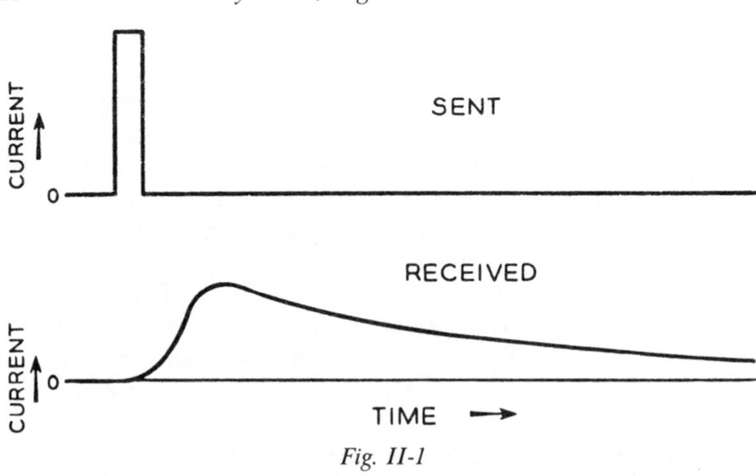

Fig. II-1

signal is transmitted it may be received as a vaguely wandering rise and fall of current which is difficult to interpret.

Of course, if we make our dots, spaces, and dashes long enough, the current at the far end will follow the current at the sending end better, but this slows the rate of transmission. It is clear that there is somehow associated with a given transmission circuit a limiting speed of transmission for dots and spaces. For submarine cables this speed is so slow as to trouble telegraphers; for wires on poles it is so fast as not to bother telegraphers. Early telegraphists were aware of this limitation, and it, too, lies at the heart of communication theory.

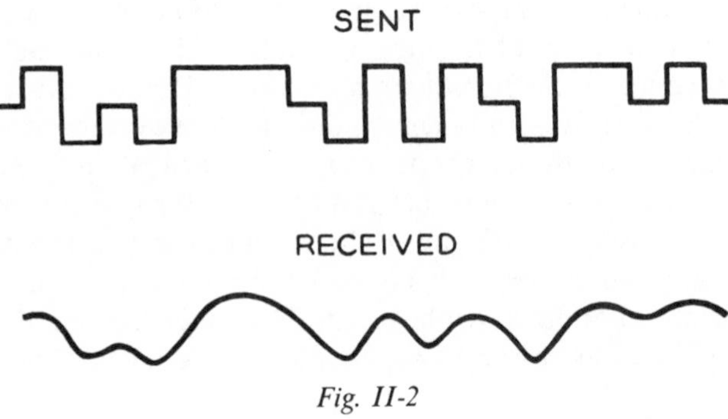

Fig. II-2

Even in the face of this limitation on speed, various things can be done to increase the number of letters which can be sent over a given circuit in a given period of time. A dash takes three times as long to send as a dot. It was soon appreciated that one could gain by means of double-current telegraphy. We can understand this by imagining that at the receiving end a galvanometer, a device which detects and indicates the direction of flow of small currents, is connected between the telegraph wire and the ground. To indicate a dot, the sender connects the positive terminal of his battery to the wire and the negative terminal to ground, and the needle of the galvanometer moves to the right. To send a dash, the sender connects the negative terminal of his battery to the wire and the positive terminal to the ground, and the needle of the galvanometer moves to the left. We say that an electric current in one direction (into the wire) represents a dot and an electric current in the other direction (out of the wire) represents a dash. No current at all (battery disconnected) represents a space. In actual double-current telegraphy, a different sort of receiving instrument is used.

In single-current telegraphy we have two elements out of which to construct our code: current and no current, which we might call 1 and 0. In double-current telegraphy we really have three elements, which we might characterize as forward current, or current into the wire; no current; backward current, or current out of the wire; or as $+1$, 0, -1. Here the $+$ or $-$ sign indicates the direction of current flow and the number 1 gives the magnitude or strength of the current, which in this case is equal for current flow in either direction.

In 1874, Thomas Edison went further; in his quadruplex telegraph system he used two intensities of current as well as two directions of current. He used changes in intensity, regardless of changes in direction of current flow to send one message, and changes of direction of current flow regardless of changes in intensity, to send another message. If we assume the currents to differ equally one from the next, we might represent the four different conditions of current flow by means of which the two messages are conveyed over the one circuit simultaneously as $+3$, $+1$, -1, -3. The interpretation of these at the receiving end is shown in Table I.

TABLE I

Current Transmitted	*Meaning*	
	Message 1	*Message* 2
+3	on	on
+1	off	on
−1	off	off
−3	on	off

Figure II-3 shows how the dots, dashes, and spaces of two simultaneous, independent messages can be represented by a succession of the four different current values.

Clearly, how much information it is possible to send over a circuit depends not only on how fast one can send successive symbols (successive current values) over the circuit but also on how many different symbols (different current values) one has available to choose among. If we have as symbols only the two currents +1 or 0 or, which is just as effective, the two currents +1 and −1, we can convey to the receiver only one of two possibilities at a time. We have seen above, however, that if we can choose among any one of four current values (any one of four symbols) at a

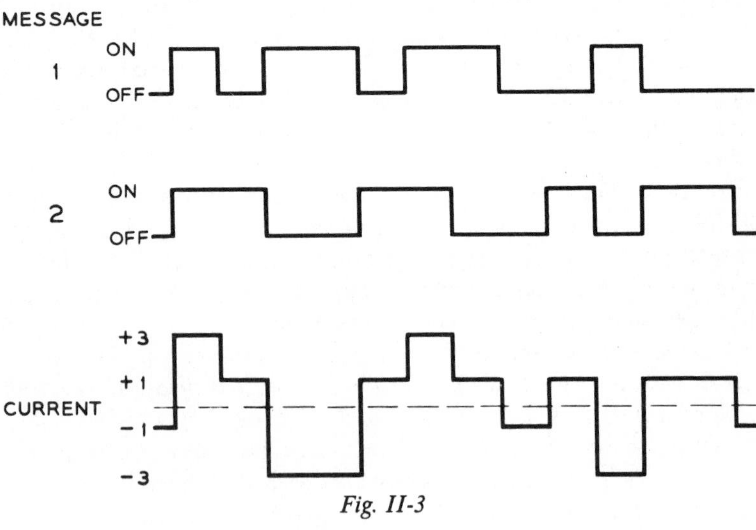

Fig. II-3

time, such as +3 or +1 or −1 or −3, we can convey by means of these current values (symbols) two independent pieces of information: whether we mean a 0 or 1 in message 1 and whether we mean a 0 or 1 in message 2. Thus, for a given rate of sending successive symbols, the use of four current values allows us to send two independent messages, each as fast as two current values allow us to send one message. We can send twice as many letters per minute by using four current values as we could using two current values.

The use of multiplicity of symbols can lead to difficulties. We have noted that dots and dashes sent over a long submarine cable tend to spread out and overlap. Thus, when we look for one symbol at the far end we see, as Figure II-2 illustrates, a little of several others. Under these circumstances, a simple identification, as 1 or 0 or else +1 or −1, is easier and more certain than a more complicated indentification, as among +3, +1, −1, −3.

Further, other matters limit our ability to make complicated distinctions. During magnetic storms, extraneous signals appear on telegraph lines and submarine cables.[2] And if we look closely enough, as we can today with sensitive electronic amplifiers, we see that minute, undesired currents are always present. These are akin to the erratic Brownian motion of tiny particles observed under a microscope and to the agitation of air molecules and of all other matter which we associate with the idea of heat and temperature. Extraneous currents, which we call *noise,* are always present to interfere with the signals sent.

Thus, even if we avoid the overlapping of dots and spaces which is called *intersymbol interference,* noise tends to distort the received signal and to make difficult a distinction among many alternative symbols. Of course, increasing the current transmitted, which means increasing the power of the transmitted signal, helps to overcome the effect of noise. There are limits on the power that can be used, however. Driving a large current through a submarine cable takes a large voltage, and a large enough voltage can destroy the insulation of the cable—can in fact cause a short circuit. It is likely that the large transmitting voltage used caused the failure of the first transatlantic telegraph cable in 1858.

[2] The changing magnetic field of the earth induces currents in the cables. The changes in the earth's magnetic field are presumably caused by streams of charged particles due to solar storms.

Even the early telegraphists understood intuitively a good deal about the limitations associated with speed of signaling, interference, or noise, the difficulty in distinguishing among many alternative values of current, and the limitation on the power that one could use. More than an intuitive understanding was required, however. An exact mathematical analysis of such problems was needed.

Mathematics was early applied to such problems, though their complete elucidation has come only in recent years. In 1855, William Thomson, later Lord Kelvin, calculated precisely what the received current will be when a dot or space is transmitted over a submarine cable. A more powerful attack on such problems followed the invention of the telephone by Alexander Graham Bell in 1875. Telephony makes use, not of the slowly sent off-on signals of telegraphy, but rather of currents whose strength varies smoothly and subtly over a wide range of amplitudes with a rapidity several hundred times as great as encountered in manual telegraphy.

Many men helped to establish an adequate mathematical treatment of the phenomena of telephony: Henri Poincaré, the great French mathematician; Oliver Heaviside, an eccentric, English, minor genius; Michael Pupin, of *From Immigrant to Inventor* fame; and G. A. Campbell, of the American Telephone and Telegraph Company, are prominent among these.

The mathematical methods which these men used were an extension of work which the French mathematician and physicist, Joseph Fourier, had done early in the nineteenth century in connection with the flow of heat. This work had been applied to the study of vibration and was a natural tool for the analysis of the behavior of electric currents which change with time in a complicated fashion—as the electric currents of telephony and telegraphy do.

It is impossible to proceed further on our way without understanding something of Fourier's contribution, a contribution which is absolutely essential to all communication and communication theory. Fortunately, the basic ideas are simple; it is their proof and the intricacies of their application which we shall have to omit here.

Fourier based his mathematical attack on some of the problems of heat flow on a very particular mathematical function called a *sine wave.* Part of a sine wave is shown at the right of Figure II-4. The height of the wave *h* varies smoothly up and down as time passes, fluctuating so forever and ever. A sine wave has no beginning or end. A sine wave is not just any smoothly wiggling curve. The height of the wave (it may represent the strength of a current or voltage) varies in a particular way with time. We can describe this variation in terms of the motion of a crank connected to a shaft which revolves at a constant speed, as shown at the left of Figure II-4. The height *h* of the crank above the axle varies exactly sinusoidally with time.

A sine wave is a rather simple sort of variation with time. It can be characterized, or described, or differentiated completely from any other sine wave by means of just three quantities. One of these is the maximum height above zero, called the *amplitude.* Another is the time at which the maximum is reached, which is specified as the *phase.* The third is the time *T* between maxima, called the *period.* Usually, we use instead of the period the reciprocal of the period called the *frequency,* denoted by the letter *f.* If the period *T* of a sine wave is 1/100 second, the frequency *f* is 100 cycles per second, abbreviated cps. A *cycle* is a complete variation from crest, through trough, and back to crest again. The sine wave is *periodic* in that one variation from crest through trough to crest again is just like any other.

Fourier succeeded in proving a theorem concerning sine waves which astonished his, at first, incredulous contemporaries. He showed that any variation of a quantity with time can be accurately represented as the sum of a number of sinusoidal variations of

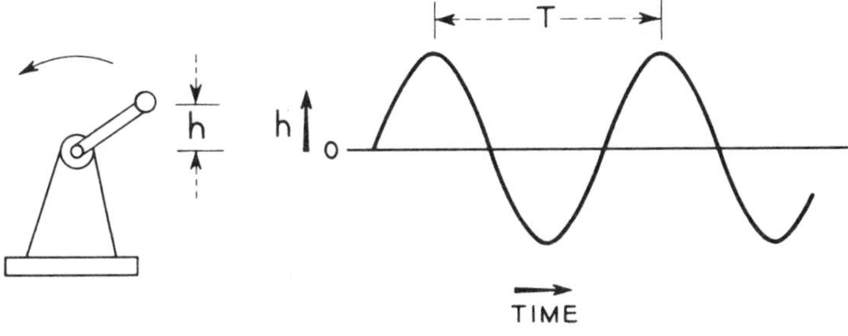

Fig. II-4

different amplitudes, phases, and frequencies. The quantity concerned might be the displacement of a vibrating string, the height of the surface of a rough ocean, the temperature of an electric iron, or the current or voltage in a telephone or telegraph wire. All are amenable to Fourier's analysis. Figure II-5 illustrates this in a simple case. The height of the periodic curve *a* above the centerline is the sum of the heights of the sinusoidal curves *b* and *c*.

The mere representation of a complicated variation of some physical quantity with time as a sum of a number of simple sinusoidal variations might seem a mere mathematician's trick. Its utility depends on two important physical facts. The circuits used in the transmission of electrical signals do not change with time, and they behave in what is called a *linear* fashion. Suppose, for instance, we send one signal, which we will call an *input signal,* over the line and draw a curve showing how the amplitude of the received signal varies with time. Suppose we send a second input signal and draw a curve showing how the corresponding received signal varies with time. Suppose we now send the sum of the two input signals, that is, a signal whose current is at every moment the simple sum of the currents of the two separate input signals. Then, the received output signal will be merely the sum of the two output signals corresponding to the input signals sent separately.

We can easily appreciate the fact that communication circuits don't change significantly with time. Linearity means simply that

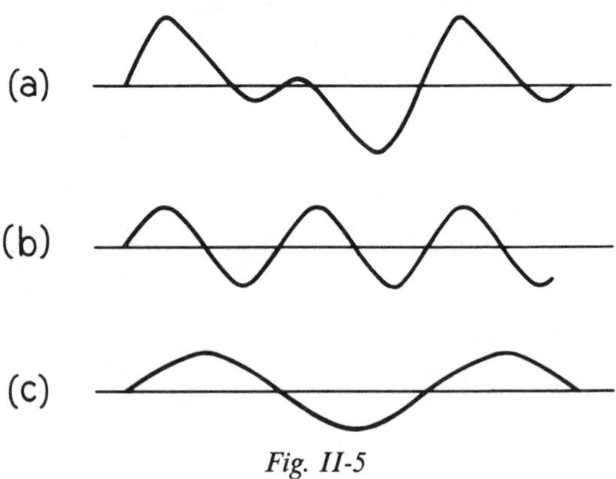

Fig. II-5

if we know the output signals corresponding to any number of input signals sent separately, we can calculate the output signal when several of the input signals are sent together merely by adding the output signals corresponding to the input signals. In a linear electrical circuit or transmission system, signals act as if they were present independently of one another; they do not interact. This is, indeed, the very criterion for a circuit being called a linear circuit.

While linearity is a truly astonishing property of nature, it is by no means a rare one. All circuits made up of the resistors, capacitors, and inductors discussed in Chapter I in connection with network theory are linear, and so are telegraph lines and cables. Indeed, usually electrical circuits are linear, except when they include vacuum tubes, or transistors, or diodes, and sometimes even such circuits are substantially linear.

Because telegraph wires are linear, which is just to say because telegraph wires are such that electrical signals on them behave independently without interacting with one another, two telegraph signals can travel in opposite directions on the same wire at the same time without interfering with one another. However, while linearity is a fairly common phenomenon in electrical circuits, it is by no means a universal natural phenomenon. Two trains can't travel in opposite directions on the same track without interference. Presumably they could, though, if all the physical phenomena comprised in trains were linear. The reader might speculate on the unhappy lot of a truly linear race of beings.

With the very surprising property of linearity in mind, let us return to the transmission of signals over electrical circuits. We have noted that the output signal corresponding to most input signals has a different shape or variation with time from the input signal. Figures II-1 and II-2 illustrate this. However, it can be shown mathematically (but not here) that, if we use a sinusoidal signal, such as that of Figure II-4, as an input signal to a linear transmission path, we always get out a sine wave of the *same* period, or frequency. The amplitude of the output sine wave may be less than that of the input sine wave; we call this *attenuation* of the sinusoidal signal. The output sine wave may rise to a peak later than the input sine wave; we call this *phase shift,* or *delay* of the sinusoidal signal.

The amounts of the attenuation and delay depend on the frequency of the sine wave. In fact, the circuit may fail entirely to transmit sine waves of some frequencies. Thus, corresponding to an input signal made up of several sinusoidal *components,* there will be an output signal having components of the same frequencies but of different relative phases or delays and of different amplitudes. Thus, in general the shape of the output signal will be different from the shape of the input signal. However, the difference can be thought of as caused by the changes in the relative delays and amplitudes of the various components, differences associated with their different frequencies. If the attenuation and delay of a circuit is the same for all frequencies, the shape of the output wave will be the same as that of the input wave; such a circuit is *distortionless.*

Because this is a very important matter, I have illustrated it in Figure II-6. In *a* we have an input signal which can be expressed as the sum of the two sinusoidal components, *b* and *c.* In transmission, *b* is neither attenuated nor delayed, so the output *b′* of the same frequency as *b* is the same as *b.* However, the output *c′* due to the input *c* is attenuated and delayed. The total output *a′,* the sum of *b′* and *c′,* clearly has a different shape from the input *a.* Yet, the output is made up of two components having the same frequencies that are present in the input. The frequency components merely have different relative phases or delays and different relative amplitudes in the output than in the input.

The *Fourier analysis* of signals into components of various frequencies makes it possible to study the transmission properties of a linear circuit for all signals in terms of the attenuation and delay it imposes on sine waves of various frequencies as they pass through it.

Fourier analysis is a powerful tool for the analysis of transmission problems. It provided mathematicians and engineers with a bewildering variety of results which they did not at first clearly understand. Thus, early telegraphists invented all sorts of shapes and combinations of signals which were alleged to have desirable properties, but they were often inept in their mathematics and wrong in their arguments. There was much dispute concerning the efficacy of various signals in ameliorating the limitations imposed by circuit speed, intersymbol interference, noise, and limitations on transmitted power.

In 1917, Harry Nyquist came to the American Telephone and Telegraph Company immediately after receiving his Ph.D. at Yale (Ph.D.'s were considerably rarer in those days). Nyquist was a much better mathematician than most men who tackled the problems of telegraphy, and he always was a clear, original, and philosophical thinker concerning communication. He tackled the problems of telegraphy with powerful methods and with clear insight. In 1924, he published his results in an important paper, "Certain Factors Affecting Telegraph Speed."

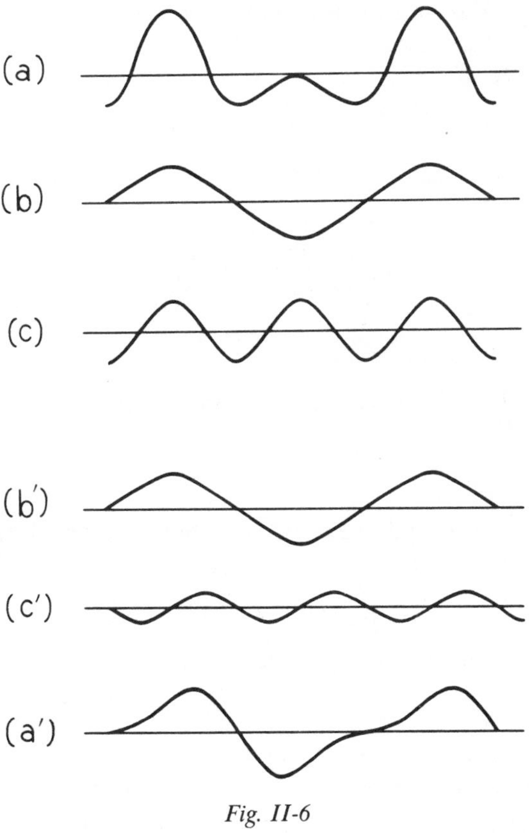

Fig. II-6

This paper deals with a number of problems of telegraphy. Among other things, it clarifies the relation between the speed of telegraphy and the number of current values such as $+1$, -1 (two current values) or $+3$, $+1$, -1, -3 (four current values). Nyquist says that if we send symbols (successive current values) at a constant rate, the speed of transmission, W, is related to m, the number of different symbols or current values available, by

$$W = \mathrm{K} \log m$$

Here K is a constant whose value depends on how many successive current values are sent each second. The quantity $\log m$ means logarithm of m. There are different *bases* for taking logarithms. If we choose 2 as a base, then the values of $\log m$ for various values of m are given in Table II.

TABLE II

m	*log m*
1	0
2	1
3	1.6
4	2
8	3
16	4

To sum up the matter by means of an equation, $\log x$ is such a number that

$$2^{\log x} = x$$

We may see by taking the logarithm of each side that the following relation must be true:

$$\log 2^{\log x} = \log x$$

If we write M in place of $\log x$, we see that

$$\log 2^M = M$$

All of this is consistent with Table II.

We can easily see by means of an example why the logarithm is the appropriate function in Nyquist's relation. Suppose that we wish to specify two independent choices of off-or-on, 0-or-1, simultaneously. There are four possible combinations of two independent 0-or-1 choices, as shown in Table III.

TABLE III

Number of Combination	*First* 0-OR-1 *Choice*	*Second* 0-OR-1 *Choice*
1	0	0
2	0	1
3	1	0
4	1	1

Further, if we wish to specify three independent choices of 0-or-1 at the same time, we find eight combinations, as shown in Table IV.

TABLE IV

Number of Combination	*First* 0-OR-1 *Choice*	*Second* 0-OR-1 *Choice*	*Third* 0-OR-1 *Choice*
1	0	0	0
2	0	0	1
3	0	1	0
4	0	1	1
5	1	0	0
6	1	0	1
7	1	1	0
8	1	1	1

Similarly, if we wish to specify four independent 0-or-1 choices, we find sixteen different combinations, and, if we wish to specify M different independent 0-or-1 choices, we find 2^M different combinations.

If we can specify M independent 0-or-1 combinations at once, we can in effect send M independent messages at once, so surely the speed should be proportional to M. But, in sending M messages at once we have 2^M possible combinations of the M independent 0-or-1 choices. Thus, to send M messages at once, we need to be able to send 2^M different symbols or current values. Suppose that we can choose among 2^M different symbols. Nyquist tells us that we should take the logarithm of the number of symbols in order to get the line speed, and

$$\log 2^M = M$$

Thus, the logarithm of the number of symbols is just the number of independent 0-or-1 choices that can be represented simultaneously, the number of independent messages we can send at once, so to speak.

Nyquist's relation says that by going from off-on telegraphy to three-current (+1, 0, −1) telegraphy we can increase the speed of sending letters or other symbols by 60 per cent, and if we use four current values (+3, +1, −1, −3) we can double the speed. This is, of course, just what Edison did with his quadruplex telegraph, for he sent two messages instead of one. Further, Nyquist showed that the use of eight current values (0, 1, 2, 3, 4, 5, 6, 7, or +7, +5, +3, +1, −1, −3, −5, −7) should enable us to send four times as fast as with two current values. However, he clearly realized that fluctuations in the attenuation of the circuit, interference or noise, and limitations on the power which can be used, make the use of many current values difficult.

Turning to the rate at which signal elements can be sent, Nyquist defined the *line speed* as one half of the number of signal elements (dots, spaces, current values) which can be transmitted in a second. We will find this definition particularly appropriate for reasons which Nyquist did not give in this early paper.

By the time that Nyquist wrote, it was common practice to send telegraph and telephone signals on the same wires. Telephony makes use of frequencies above 150 cps, while telegraphy can be carried out by means of lower frequency signals. Nyquist showed how telegraph signals could be so shaped as to have no sinusoidal components of high enough frequency to be heard as interference by telephones connected to the same line. He noted that the line speed, and hence also the speed of transmission, was proportional to the width or extent of the range or *band* (in the sense of strip) of frequencies used in telegraphy; we now call this range of frequencies the *band width* of a circuit or of a signal.

Finally, in analyzing one proposed sort of telegraph signal,

Nyquist showed that it contained at all times a steady sinusoidal component of constant amplitude. While this component formed a part of the transmitter power used, it was useless at the receiver, for its eternal, regular fluctuations were perfectly predictable and could have been supplied at the receiver rather than transmitted thence over the circuit. Nyquist referred to this useless component of the signal, which, he said, conveyed no intelligence, as *redundant,* a word which we will encounter later.

Nyquist continued to study the problems of telegraphy, and in 1928 he published a second important paper, "Certain Topics in Telegraph Transmission Theory." In this he demonstrated a number of very important points. He showed that if one sends some number $2N$ of different current values per second, all the sinusoidal components of the signal with frequencies greater than N are redundant, in the sense that they are not needed in deducing from the received signal the succession of current values which were sent. If all of these higher frequencies were removed, one could still deduce by studying the signal which current values had been transmitted. Further, he showed how a signal could be constructed which would contain no frequencies above N cps and from which it would be very easy to deduce at the receiving point what current values had been sent. This second paper was more quantitative and exact than the first; together, they embrace much important material that is now embodied in communication theory.

R. V. L. Hartley, the inventor of the Hartley oscillator, was thinking philosophically about the transmission of information at about this time, and he summarized his reflections in a paper, "Transmission of Information," which he published in 1928.

Hartley had an interesting way of formulating the problem of communication, one of those ways of putting things which may seem obvious when stated but which can wait years for the insight that enables someone to make the statement. He regarded the sender of a message as equipped with a set of symbols (the letters of the alphabet for instance) from which he mentally selects symbol after symbol, thus generating a sequence of symbols. He observed that a chance event, such as the rolling of balls into pockets, might equally well generate such a sequence. He then defined H, the information of the message, as the logarithm of the number of possible sequences of symbols which might have been selected and showed that

$$H = n \log s$$

Here n is the number of symbols selected, and s is the number of different symbols in the set from which symbols are selected.

This is acceptable in the light of our present knowledge of information theory only if successive symbols are chosen independently and if any of the s symbols is equally likely to be selected. In this case, we need merely note, as before, that the logarithm of s, the number of symbols, is the number of independent 0-or-1 choices that can be represented or sent simultaneously, and it is reasonable that the rate of transmission of information should be the rate of sending symbols per second n, times the number of independent 0-or-1 choices that can be conveyed per symbol.

Hartley goes on to the problem of encoding the primary symbols (letters of the alphabet, for instance) in terms of secondary symbols (e.g., the sequences of dots, spaces, and dashes of the Morse code). He observes that restrictions on the selection of symbols (the fact that E is selected more often than Z) should govern the lengths of the secondary symbols (Morse code representations) if we are to transmit messages most swiftly. As we have seen, Morse himself understood this, but Hartley stated the matter in a way which encouraged mathematical attack and inspired further work. Hartley also suggested a way of applying such considerations to continuous signals, such as telephone signals or picture signals.

Finally, Hartley stated, in accord with Nyquist, that the amount of information which can be transmitted is proportional to the band width times the time of transmission. But this makes us wonder about the number of allowable current values, which is also important to speed of transmission. How are we to enumerate them?

After the work of Nyquist and Hartley, communication theory appears to have taken a prolonged and comfortable rest. Workers busily built and studied particular communication systems. The art grew very complicated indeed during World War II. Much new understanding of particular new communication systems and devices was achieved, but no broad philosophical principles were laid down.

During the war it became important to predict from inaccurate or "noisy" radar data the courses of airplanes, so that the planes could be shot down. This raised an important question: Suppose that one has a varying electric current which represents data concerning the present position of an airplane but that there is added to it a second meaningless erratic current, that is, a noise. It may be that the frequencies most strongly present in the signal are different from the frequencies most strongly present in the noise. If this is so, it would seem desirable to pass the signal with the noise added through an electrical circuit or *filter* which attenuates the frequencies strongly present in the noise but does not attenuate very much the frequencies strongly present in the signal. Then, the resulting electric current can be passed through other circuits in an effort to estimate or predict what the value of the original signal, without noise, will be a few seconds from the present. But what sort of combination of electrical circuits will enable one best to predict from the present noisy signal the value of the true signal a few seconds in the future?

In essence, the problem is one in which we deal with not one but with a whole *ensemble* of possible signals (courses of the plane), so that we do not know in advance which signal we are dealing with. Further, we are troubled with an unpredictable noise.

This problem was solved in Russia by A. N. Kolmogoroff. In this country it was solved independently by Norbert Wiener. Wiener is a mathematician whose background ideally fitted him to deal with this sort of problem, and during the war he produced a yellow-bound document, affectionately called "the yellow peril" (because of the headaches it caused), in which he solved the difficult problem.

During and after the war another mathematician, Claude E. Shannon, interested himself in the general problem of communication. Shannon began by considering the relative advantages of many new and fanciful communication systems, and he sought some basic method of comparing their merits. In the same year (1948) that Wiener published his book, *Cybernetics,* which deals with communication and control, Shannon published in two parts a paper which is regarded as the foundation of modern communication theory.

Wiener and Shannon alike consider, not the problem of a single signal, but the problem of dealing adequately with *any* signal selected from a group or ensemble of possible signals. There was a free interchange among various workers before the publication of either Wiener's book or Shannon's paper, and similar ideas and expressions appear in both, although Shannon's interpretation appears to be unique.

Chiefly, Wiener's name has come to be associated with the field of extracting signals of a given ensemble from noise of a known type. An example of this has been given above. The enemy pilot follows a course which he choses, and our radar adds noise of natural origin to the signals which represent the position of the plane. We have a set of possible signals (possible courses of the airplane), not of our own choosing, mixed with noise, not of our own choosing, and we try to make the best estimate of the present or future value of the signal (the present or future position of the airplane) despite the noise.

Shannon's name has come to be associated with matters of so encoding messages chosen from a known ensemble that they can be transmitted accurately and swiftly in the presence of noise. As an example, we may have as a message source English text, not of our own choosing, and an electrical circuit, say, a noisy telegraph cable, not of our own choosing. But in the problem treated by Shannon, we are allowed to choose how we shall represent the message as an electrical signal—how many current values we shall allow, for instance, and how many we shall transmit per second. The problem, then, is not how to treat a signal plus noise so as to get a best estimate of the signal, but what sort of signal to send so as best to convey messages of a given type over a particular sort of noisy circuit.

This matter of efficient encoding and its consequences form the chief substance of information theory. In that an ensemble of messages is considered, the work reflects the spirit of the work of Kolmogoroff and Wiener and of the work of Morse and Hartley as well.

It would be useless to review here the content of Shannon's

work, for that is what this book is about. We shall see, however, that it sheds further light on all the problems raised by Nyquist and Hartley and goes far beyond those problems.

In looking back on the origins of communication theory, two other names should perhaps be mentioned. In 1946, Dennis Gabor published an ingenious paper, "Theory of Communication." This, suggestive as it is, missed the inclusion of noise, which is at the heart of modern communication theory. Further, in 1949, W. G. Tuller published an interesting paper, "Theoretical Limits on the Rate of Transmission of Information," which in part parallels Shannon's work.

The gist of this chapter has been that the very general theory of communication which Shannon has given us grew out of the study of particular problems of electrical communication. Morse was faced with the problem of representing the letters of the alphabet by short or long pulses of current with intervening spaces of no current—that is, by the dots, dashes, and spaces of telegraphy. He wisely chose to represent common letters by short combinations of dots and dashes and uncommon letters by long combinations; this was a first step in efficient encoding of messages, a vital part of communication theory.

Ingenious inventors who followed Morse made use of different intensities and directions of current flow in order to give the sender a greater choice of signals than merely off-or-on. This made it possible to send more letters per unit time, but it made the signal more susceptible to disturbance by unwanted electrical disturbances called noise as well as by inability of circuits to transmit accurately rapid changes of current.

An evaluation of the relative advantages of many different sorts of telegraph signals was desirable. Mathematical tools were needed for such a study. One of the most important of these is Fourier analysis, which makes it possible to represent any signal as a sum of sine waves of various frequencies.

Most communication circuits are linear. This means that several signals present in the circuit do not interact or interfere. It can be shown that while even linear circuits change the shape of most signals, the effect of a linear circuit on a sine wave is merely to make it weaker and to delay its time of arrival. Hence, when a complicated signal is represented as a sum of sine waves of various frequencies, it is easy to calculate the effect of a linear circuit on each sinusoidal component separately and then to add up the weakened or attenuated sinusoidal components in order to obtain the over-all received signal.

Nyquist showed that the number of distinct, different current values which can be sent over a circuit per second is twice the total range or band width of frequencies used. Thus, the rate at which letters of text can be transmitted is proportional to band width. Nyquist and Hartley also showed that the rate at which letters of text can be transmitted is proportional to the logarithm of the number of current values used.

A complete theory of communication required other mathematical tools and new ideas. These are related to work done by Kolmogoroff and Wiener, who considered the problem of an unknown signal of a given type disturbed by the addition of noise. How does one best estimate what the signal is despite the presence of the interfering noise? Kolmogoroff and Wiener solved this problem.

The problem Shannon set himself is somewhat different. Suppose we have a message source which produces messages of a given type, such as English text. Suppose we have a noisy communication channel of specified characteristics. How can we represent or encode messages from the message source by means of electrical signals so as to attain the fastest possible transmission over the noisy channel? Indeed, how fast can we transmit a given type of message over a given channel without error? In a rough and general way, this is the problem that Shannon set himself and solved.

CHAPTER III *A Mathematical Model*

A MATHEMATICAL THEORY which seeks to explain and to predict the events in the world about us always deals with a simplified model of the world, a mathematical model in which only things pertinent to the behavior under consideration enter.

Thus, planets are composed of various substances, solid, liquid, and gaseous, at various pressures and temperatures. The parts of their substances exposed to the rays of the sun reflect various fractions of the different colors of the light which falls upon them, so that when we observe planets we see on them various colored features. However, the mathematical astronomer in predicting the orbit of a planet about the sun need take into account only the total mass of the sun, the distance of the planet from the sun, and the speed and direction of the planet's motion at some initial instant. For a more refined calculation, the astronomer must also take into account the total mass of the planet and the motions and masses of other planets which exert gravitational forces on it.

This does not mean that astronomers are not concerned with other aspects of planets, and of stars and nebulae as well. The important point is that they need not take these other matters into consideration in computing planetary orbits. The great beauty and power of a mathematical theory or model lies in the separation of the relevant from the irrelevant, so that certain observable behavior can be related and understood without the need of comprehending the whole nature and behavior of the universe.

Mathematical models can have various degrees of accuracy or applicability. Thus, we can accurately predict the orbits of planets by regarding them as rigid bodies, despite the fact that no truly rigid body exists. On the other hand, the long-term motions of our moon can only be understood by taking into account the motion of the waters over the face of the earth, that is, the tides. Thus, in dealing very precisely with lunar motion we cannot regard the earth as a rigid body.

In a similar way, in network theory we study the electrical properties of interconnections of ideal inductors, capacitors, and resistors, which are assigned certain simple mathematical properties. The components of which the actual useful circuits in radio, TV, and telephone equipment are made only approximate the properties of the ideal inductors, capacitors, and resistors of network theory. Sometimes, the difference is trivial and can be disregarded. Sometimes it must be taken into account by more refined calculations.

Of course, a mathematical model may be a very crude or even an invalid representation of events in the real world. Thus, the self-interested, gain-motivated "economic man" of early economic theory has fallen into disfavor because the behavior of the economic man does not appear to correspond to or to usefully explain the actual behavior of our economic world and of the people in it.

In the orbits of the planets and the behavior of networks, we have examples of idealized *deterministic* systems which have the sort of predictable behavior we ordinarily expect of machines. Astronomers can compute the positions which the planets will occupy millennia in the future. Network theory tells us all the subsequent behavior of an electrical network when it is excited by a particular electrical signal.

Even the individual economic man is deterministic, for he will always act for his economic gain. But, if he at some time gambles on the honest throw of a die because the odds favor him, his economic fate becomes to a degree unpredictable, for he may lose even though the odds do favor him.

We can, however, make a mathematical model for purely chance events, such as the drawing of some number, say three, of white or black balls from a container holding equal numbers of white and black balls. This model tells us, in fact, that after many trials we will have drawn all white about ⅛ of the time, two whites and a black about ⅜ of the time, two blacks and a white about ⅜ of the time, and all black about ⅛ of the time. It can also tell us how much of a deviation from these proportions we may reasonably expect after a given number of trials.

Our experience indicates that the behavior of actual human beings is neither as determined as that of the economic man nor as simply random as the throw of a die or as the drawing of balls from a mixture of black and white balls. It is clear, however, that a deterministic model will not get us far in the consideration of human behavior, such as human communication, while a random or statistical model might.

We all know that the actuarial tables used by insurance companies make fair predictions of the fraction of a large group of men in a given age group who will die in one year, despite the fact that we cannot predict when a particular man will die. Thus a statistical model may enable us to understand and even to make some sort of predictions concerning human behavior, even as we can predict how often, on the average, we will draw three black balls by chance from an equal mixture of white and black balls.

It might be objected that actuarial tables make predictions concerning groups of people, not predictions concerning individuals. However, experience teaches us that we can make predictions concerning the behavior of *individual* human beings as well as of groups of individuals. For instance, in counting the frequency of usage of the letter E in all English prose we will find that E constitutes about 0.13 of all the letters appearing, while W, for instance, constitutes only about 0.02 of all letters appearing. But, we also find almost the same proportions of E's and W's in the prose written by any one person. Thus, we can predict with some confidence that if you, or I, or Joe Doakes, or anyone else writes a long letter, or an article, or a book, about 0.13 of the letters he uses will be E's.

This predictability of behavior limits our freedom no more than does any other habit. We don't have to use in our writing the same fraction of E's, or of any other letter, that everyone else does. In fact, several untrammeled individuals have broken away from the common pattern. William F. Friedman, the eminent cryptanalyst and author of *The Shakesperian Cipher Examined,* has supplied me with the following examples.

Gottlob Burmann, a German poet who lived from 1737 to 1805, wrote 130 poems, including a total of 20,000 words, without once using the letter R. Further, during the last seventeen years of his life, Burmann even omitted the letter from his daily conversation.

In each of five stories published by Alonso Alcala y Herrera in Lisbon in 1641 a different vowel was suppressed. Francisco Navarrete y Ribera (1659), Fernando Jacinto de Zurita y Haro (1654), and Manuel Lorenzo de Lizarazu y Berbuizana (1654) provided other examples.

In 1939, Ernest Vincent Wright published a 267-page novel, *Gadsby,* in which no use is made of the letter E. I quote a paragraph below:

> Upon this basis I am going to show you how a bunch of bright young folks did find a champion; a man with boys and girls of his own; a man of so dominating and happy individuality that Youth is drawn to him as is a fly to a sugar bowl. It is a story about a small town. It is not a gossipy yarn; nor is it a dry, monotonous account, full of such customary "fill-ins" as "romantic moonlight casting murky shadows down a long, winding country road." Nor will it say anything about tinklings lulling distant folds; robins carolling at twilight, nor any "warm glow of lamplight" from a cabin window. No. It is an account of up-and-doing activity; a vivid portrayal of Youth as it is today; and a practical discarding of that worn-out notion that "a child don't know anything."

While such exercises of free will show that it is not impossible to break the chains of habit, we ordinarily write in a more conventional manner. When we are not going out of our way to demonstrate that we can do otherwise, we customarily use our due fraction of 0.13 E's with almost the consistency of a machine or a mathematical rule.

We cannot argue from this to the converse idea that a machine into which the same habits were built could write English text. However, Shannon has demonstrated how English words and text can be approximated by a mathematical process which could be carried out by a machine.

Suppose, for instance, that we merely produce a sequence of letters and spaces with equal probabilities. We might do this by putting equal numbers of cards marked with each letter and with the space into a hat, mixing them up, drawing a card, recording its symbol, returning it, remixing, drawing another card, and so on. This gives what Shannon calls the zero-order approximation to English text. His example, obtained by an equivalent process, goes:

1. Zero-order approximation (symbols independent and equiprobable)

XFOML RXKHRJFFJUJ ZLPWCFWKCYJ FFJEYVKCQSGHYD QPAAMKBZAACIBZLHJQD.

Here there are far too many Zs and Ws, and not nearly enough E's and spaces. We can approach more nearly to English text by choosing letters independently of one another, but choosing E more often than W or Z. We could do this by putting many E's and few W's and Z's into the hat, mixing, and drawing out the letters. As the *probability* that a given letter is an E should be .13, out of every hundred letters we put into the hat, 13 should be E's. As the probability that a letter will be W should be .02, out of each hundred letters we put into the hat, 2 should be W's, and so on. Here is the result of an equivalent procedure, which gives what Shannon calls a first-order approximation of English text:

2. First-order approximation (symbols independent but with frequencies of English text).

OCRO HLI RGWR NMIELWIS EU LL NBNESEBYA TH EEI ALHENHTTPA OOBTTVA NAH BRL

In English text we almost never encounter any pair of letters beginning with Q except QU. The probability of encountering QX or QZ is essentially zero. While the probability of QU is not 0, it is so small as not to be listed in the tables I consulted. On the other hand, the probability of TH is .037, the probability of OR is .010 and the probability of WE is .006. These probabilities have the following meaning. In a stretch of text containing, say, 10,001 letters, there are 10,000 successive pairs of letters, i.e., the first and second, the second and third, and so on to the next to last and the last. Of the pairs a certain number are the letters TH. This might be 370 pairs. If we divide the total number of times we find TH, which we have assumed to be 370 times, by the total number of pairs of letters, which we have assumed to be 10,000, we get the probability that a randomly selected pair of letters in the text will be TH, that is, 370/10,000, or .037.

Diligent cryptanalysts have made tables of such *digram probabilities* for English text. To see how we might use these in constructing sequences of letters with the same digram probabilities as English text, let us assume that we use 27 hats, 26 for digrams beginning with each of the letters and one for digrams beginning with a space. We will then put a large number of digrams into the hats according to the probabilities of the digrams. Out of 1,000 digrams we would put in 37 TH's, 10 WE's, and so on.

Let us consider for a moment the meaning of these hats full of digrams in terms of the original counts which led to the evaluations of digram probabilities.

In going through the text letter by letter we will encounter every T in the text. Thus, the number of digrams *beginning* with T, all of which we put in one hat, will be the same as the number of T's. The fraction these represent of the total number of digrams counted is the probability of encountering T in the text; that is, .10. We might call this probability $p(\mathrm{T})$

$$p(\mathrm{T}) = .10$$

We may note that this is also the fraction of digrams, distributed among the hats, which *end* in T as well as the fraction that *begin* with T.

Again, basing our total numbers on 1,001 letters of text, or 1,000 digrams, the number of times the digram TH is encountered is 37, and so the probability of encountering the digram TH, which we might call $p(\mathrm{T}, \mathrm{H})$ is

$$p(\mathrm{T}, \mathrm{H}) = .037$$

Now we see that 0.10, or 100, of the digrams will begin with T and hence will be in the T hat and of these 37 will be TH. Thus, the fraction of the T digrams which are TH will be 37/100, or 0.37. Correspondingly, we say that the probability that a digram beginning with T is TH, which we might call $p_{\mathrm{T}}(\mathrm{H})$, is

$$p_{\mathrm{T}}(\mathrm{H}) = .37$$

This is called the *conditional probability* that the letter following a T will be an H.

One can use these probabilities, which are adequately represented by the numbers of various digrams in the various hats, in the construction of text which has both the same *letter* frequencies and *digram* frequencies as does English text. To do this one draws the first digram at random from any hat and writes down its letters. He then draws a second digram from the hat indicated by the second letter of the first digram and writes down the second letter of this second digram. Then he draws a third digram from the hat indicated by the second letter of the second digram and writes down the second letter of this third digram, and so on. The space is treated just like a letter. There is a particular probability that a space will follow a particular letter (ending a "word") and a particular probability that a particular letter will follow a space (starting a new "word").

By an equivalent process, Shannon constructed what he calls a second-order approximation to English; it is:

3. Second-order approximation (digram structure as in English).

ON IE ANTSOUTINYS ARE T INCTORE ST BE S DEAMY ACHIN D ILONASIVE TUCOOWE AT TEASONARE FUSO TIZIN ANDY TOBE SEACE CTISBE

Cryptanalysts have even produced tables giving the probabilities of groups of three letters, called *trigram probabilities.* These can be used to construct what Shannon calls a third-order approximation to English. His example goes:

4. Third-order approximation (trigram structure as in English).

IN NO IST LAT WHEY CRATICT FROURE BIRS GROCID PONDENOME OF DEMONSTURES OF THE REPTAGIN IS REGOACTIONA OF CRE

When we examine Shannon's examples 1 through 4 we see an increasing resemblance to English text. Example 1, the zero-order approximation, has no wordlike combinations. In example 2, which takes letter frequencies into account, OCRO and NAH somewhat resemble English words. In example 3, which takes digram frequencies into account, all the "words" are pronounceable, and ON, ARE, BE, AT, and ANDY occur in English. In example 4, which takes trigram frequencies into account, we have eight English words and many English-sounding words, such as GROCID, PONDENOME, and DEMONSTURES.

G. T. Guilbaud has carried out a similar process using the statistics of Latin and has so produced a third-order approximation (one taking into account trigram frequencies) resembling Latin, which I quote below:

> IBUS CENT IPITIA VETIS <u>IPSE</u> <u>CUM</u> VIVIVS <u>SE</u> ACETITI DEDENTUR

The underlined words are genuine Latin words.

It is clear from such examples that by giving a machine certain statistics of a language, the probabilities of finding a particular letter or group of 1, or 2, or 3, or *n* letters, and by giving the machine an ability equivalent to picking a ball from a hat, flipping a coin, or choosing a random number, we could make the machine produce a close approximation to English text or to text in some other language. The more complete information we gave the machine, the more closely would its product resemble English or other text, both in its statistical structure and to the human eye.

If we allow the machine to choose groups of three letters on the basis of their probability, then any three-letter combination which it produces must be an English word or a part of an English word and any two letter "word" must be an English word. The machine is, however, less inhibited than a person, who ordinarily writes down only sequences of letters which do spell words. Thus, he misses ever writing down pompous PONDENOME, suspect ILONASIVE, somewhat vulgar GROCID, learned DEMONSTURES, and wacky but delightful DEAMY. Of course, a man in principle *could* write down such combinations of letters but ordinarily he doesn't.

We could cure the machine of this ability to produce un-English words by making it choose among groups of letters as long as the longest English word. But, it would be much simpler merely to supply the machine with words rather than letters and to let it produce these words according to certain probabilities.

Shannon has given an example in which words were selected independently, but with the probabilities of their occurring in English text, so that *the, and, man,* etc., occur in the same proportion as in English. This could be achieved by cutting text into words, scrambling the words in a hat, and then drawing out a succession of words. He calls this a first-order word approximation. It runs as follows:

5. First-order word approximation. Here words are chosen independently but with their appropriate frequencies.

REPRESENTING AND SPEEDILY IS AN GOOD APT OR COME CAN DIFFERENT NATURAL HERE HE THE A IN CAME THE TO OF TO EXPERT GRAY COME TO FURNISHES THE LINE MESSAGE HAD BE THESE

There are no tables which give the probability of different pairs of words. However, Shannon constructed a random passage in which the probabilities of pairs of words were the same as in English text by the following expedient. He chose a first pair of words at random in a novel. He then looked through the novel for the next occurrence of the second word of the first pair and added the word which followed it in this new occurrence, and so on.

This process gave him the following second-order word approximation to English.

6. Second-order word approximation. The word transition probabilities are correct, but no further structure is included.

THE HEAD AND IN FRONTAL ATTACK ON AN ENGLISH WRITER THAT THE CHARACTER OF THIS POINT IS THEREFORE ANOTHER METHOD FOR THE LETTERS THAT THE TIME OF WHO EVER TOLD THE PROBLEM FOR AN UNEXPECTED.

We see that there are stretches of several words in this final passage which resemble and, indeed, might occur in English text.

Let us consider what we have found. In actual English text, in that text which we send by teletypewriter, for instance, particular letters occur with very nearly constant frequencies. Pairs of letters and triplets and quadruplets of letters occur with almost constant frequencies over long stretches of the text. Words and pairs of words occur with almost constant frequencies. Further, we can by means of a random mathematical process, carried out by a machine if you like, produce sequences of English words or letters exhibiting these same statistics.

Such a scheme, even if refined greatly, would not, however, produce all sequences of words that a person might utter. Carried to an extreme, it would be confined to combinations of words which *had* occurred; otherwise, there would be no statistical data available on them. Yet I may say, "The magenta typhoon whirled the farded bishop away," and this may well never have been said before.

The real rules of English text deal not with letters or words alone but with classes of words and their rules of association, that is, with grammar. Linguists and engineers who try to make machines for translating one language into another must find these rules, so that their machines can combine words to form grammatical utterances even when these exact combinations have not occurred before (and also so that the meaning of words in the text to be translated can be deduced from the context). This is a big problem. It is easy, however, to describe a "machine" which randomly produces endless, grammatical utterances of a limited sort.

Figure III-1 is a diagram of such a "machine." Each numbered box represents a *state* of the machine. Because there is only a finite number of boxes or states, this is called a *finite-state* machine.

From each box a number of arrows go to other boxes. In this particular machine, only two arrows go from each box to each of two other boxes. Also, in this case, each arrow is labeled ½. This indicates that the probability of the machine passing from, for instance, state 2 to state 3 is ½ and the probability of the machine passing from state 2 to state 4 is ½.

To make the machine run, we need a sequence of random choices, which we can obtain by flipping a coin repeatedly. We can let *heads* (H) mean *follow the top arrow* and *tails* (T), *follow the bottom arrow*. This will tell us to pass to a new state. When we do this we print out the word, words, or symbol written in that state box and flip again to get a new state.

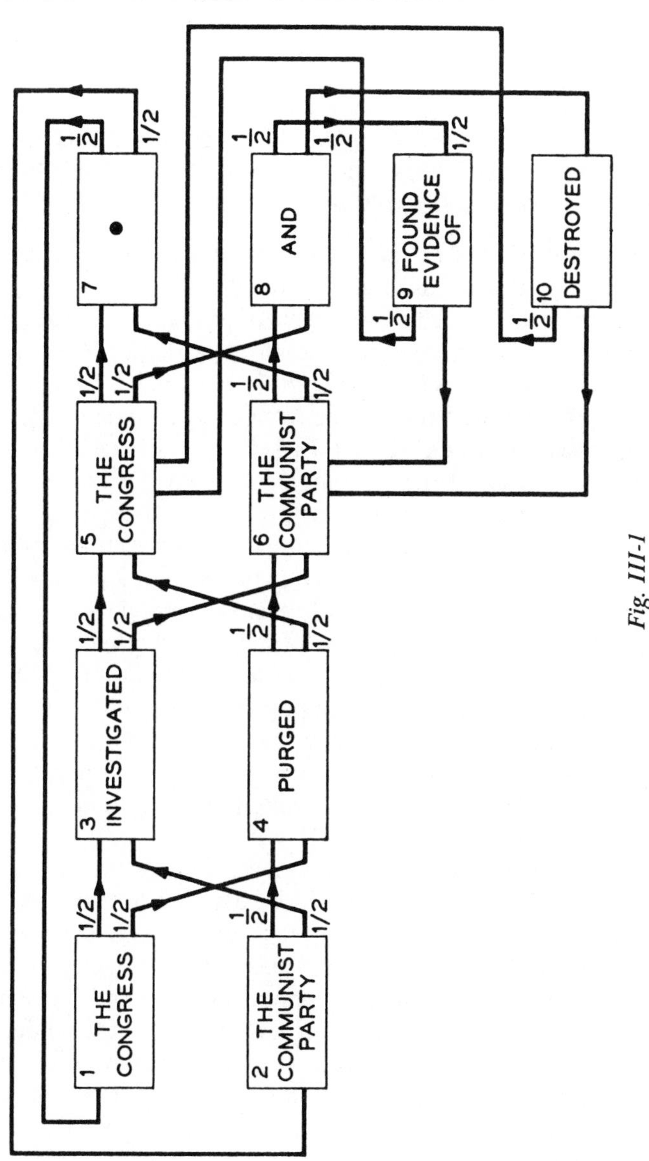
1 THE CONGRESS
2 THE COMMUNIST PARTY
3 INVESTIGATED
4 PURGED
5 THE CONGRESS
6 THE COMMUNIST PARTY
7
8 AND
9 FOUND EVIDENCE OF
10 DESTROYED
1/2

Fig. III-1

As an example, if we started in state 7 and flipped the following sequence of heads and tails: T H H H T T H T T T H H H H, the "machine would print out"

THE COMMUNIST PARTY INVESTIGATED THE CONGRESS. THE COMMUNIST PARTY PURGED THE CONGRESS AND DESTROYED THE COMMUNIST PARTY AND FOUND EVIDENCE OF THE CONGRESS.

This can go on and on, never retracing its whole course and producing "sentences" of unlimited length.

Random choice according to a table of probabilities of sequences of symbols (letters and space) or words can produce material resembling English text. A finite-state machine with a random choice among allowed transitions from state to state can produce material resembling English text. Either process is called a *stochastic* process, because of the random element involved in it.

We have examined a number of properties of English text. We have seen that the average frequency of E's is commonly constant for both the English text produced by one writer and, also, for the text produced by all writers. Other more complicated statistics, such as the frequency of digrams (TH, WE, and other letter pairs), are also essentially constant. Further, we have shown that English-like text can be produced by a sequence of random choices, such as drawings of slips of paper from hats, or flips of a coin, if the proper probabilities are in some way built into the process. One way of producing such text is through the use of a finite-state machine, such as that of Figure III-1.

We have been seeking a mathematical model of a source of English text. Such a model should be capable of producing text which corresponds closely to actual English text, closely enough so that the problem of encoding and transmitting such text is essentially equivalent to the problem of encoding and transmitting actual English text. The mathematical properties of the model must be mathematically defined so that useful theorems can be proved concerning the encoding and transmission of the text is produces, theorems which are applicable to a high degree of approximation to the encoding of actual English text. It would, however, be asking too much to insist that the production of actual English text conform with mathematical exactitude to the operation of the model.

The mathematical model which Shannon adopted to represent the production of text (and of spoken and visual messages as well) is the *ergodic source.* To understand what an ergodic source is, we must first understand what a *stationary source* is, and to explain this is our next order of business.

The general idea of a stationary source is well conveyed by the name. Imagine, for instance, a process, i.e., an imaginary machine, that produces forever after it is started the sequences of characters

A E A E A E A E A E, etc.

Clearly, what comes later is like what has gone before, and *stationary* seems an apt designation of such a source of characters. We might contrast this with a source of characters which, after starting, produced

A E A A E E A A A E E E, etc.

Here the strings of A's and E's get longer and longer without end; certainly this is not a stationary source.

Similarly, a sequence of characters chosen at random with some assigned probabilities (the first-order letter approximation of example 1 above) constitutes a stationary source and so do the digram and trigram sources of examples 2 and 3. The general idea of a stationary source is clear enough. An adequate mathematical definition is a little more difficult.

The idea of stationarity of a source demands no change with time. Yet, consider a digram source, in which the probability of the second character depends on what the previous character is. If we start such a source out on the letter A, several different letters can follow, while if we start such a source out on the letter Q, the second letter must be U. In general, the manner of starting the source will influence the statistics of the sequence of characters produced, at least for some distance from the start.

To get around this, the mathematician says, let us not consider just one sequence of characters produced by the source. After all, our source is an imaginary machine, and we can quite well imagine that it has been started an infinite number of times, so as to produce an infinite number of sequences of characters. Such an infinite number of sequences is called an *ensemble* of sequences.

These sequences could be started in any specified manner. Thus, in the case of a digram source, we can if we wish start a fraction, 0.13, of the sequences with E (this is just the probability of E in English text), a fraction, 0.02, with W (the probability of W), and so on. *If we do this,* we will find that the fraction of E's is the same, averaging over all the *first* letters of the ensemble of sequences, as it is averaging over all the *second* letters of the ensemble, as it is averaging over all the *third* letters of the ensemble, and so on. No matter what position from the beginning we choose, the fraction of E's or of any other letter occurring in that position, taken over all the sequences in the ensemble, is the same. This independence with respect to position will be true also for the probability with which TH or WE occurs among the first, second, third, and subsequent *pairs* of letters in the sequences of the ensemble.

This is what we mean by stationarity. If we can find a way of assigning probabilities to the various starting conditions used in forming the ensemble of sequences of characters which we allow the source to produce, probabilities such that any statistic obtained by averaging over the ensemble doesn't depend on the distance from the start at which we take an average, then the source is said to be stationary. This may seem difficult or obscure to the reader, but the difficulty arises in giving a useful and exact mathematical form to an idea which would otherwise be mathematically useless.

In the argument above we have, in discussing the infinite ensemble of sequences produced by a source, considered averaging over-all *first* characters or over-all *second* or *third* characters (or pairs, or triples of characters, as other examples). Such an average is called an *ensemble* average. It is different from a sort of average we talked about earlier in this chapter, in which we lumped together all the characters in *one* sequence and took the average over them. Such an average is called a *time* average.

The time average and the ensemble average can be different. For instance, consider a source which starts a third of the time with A and produces alternately A and B, a third of the time with B and produces alternately B and A, and a third of the time with E and produces a string of E's. The possible sequences are

1. A B A B A B A B, etc.
2. B A B A B A B A, etc.
3. E E E E E E E E, etc.

We can see that this is a stationary source, yet we have the probabilities shown in Table V.

TABLE V

Probability of	*Time Average Sequence (1)*	*Time Average Sequence (2)*	*Time Average Sequence (3)*	*Ensemble Average*
A	½	½	0	⅓
B	½	½	0	⅓
E	0	0	1	⅓

When a source is stationary, and when every possible ensemble average (of letters, digrams, trigrams, etc.) is equal to the corresponding time average, the source is said to be ergodic. The theorems of information theory which are discussed in subsequent chapters apply to ergodic sources, and their proofs rest on the assumption that the message source is ergodic.[1]

While we have here discussed *discrete* sources which produce sequences of characters, information theory also deals with continuous sources, which generate smoothly varying signals, such as the acoustic waves of speech or the fluctuating electric currents which correspond to these in telephony. The sources of such signals are also assumed to be ergodic.

Why is an ergodic message source an appropriate and profitable mathematical model for study? For one thing, we see by examining the definition of an ergodic source as given above that for an ergodic source the statistics of a message, for instance, the frequency of occurrence of a letter, such as E, or of a digram, such as TH, do not vary along the length of the message. As we analyze a longer and longer stretch of a message, we get a better and better estimate of the probabilities of occurrence of various letters and letter groups. In other words, by examining a longer and longer stretch of a message we are able to arrive at and refine a mathematical description of the source.

Further, the probabilities, the description of the source arrived at through such an examination of one message, apply equally well to *all* messages generated by the source and not just to the

[1] Some work has been done on the encoding of nonstationary sources, but it is not discussed in this book.

particular message examined. This is assured by the fact that the time and ensemble averages are the same.

Thus, an ergodic source is a particularly simple kind of probabilistic or stochastic source of messages, and simple processes are easier to deal with mathematically than are complicated processes. However, simplicity in itself is not enough. The ergodic source would not be of interest in communication theory if it were not reasonably realistic as well as simple.

Communication theory has two sides. It has a mathematically exact side, which deals rigorously with hypothetical, exactly ergodic sources, sources which we can imagine to produce infinite ensembles of infinite sequences of symbols. Mathematically, we are free to investigate rigorously either such a source itself or the infinite ensemble of messages which it can produce.

We *use* the theorems of communication theory in connection with the transmission of actual English text. A human being is not a hypothetical, mathematically defined machine. He cannot produce even one infinite sequence of characters, let alone an infinite ensemble of sequences.

A man does, however, produce many long sequences of characters, and all the writers of English together collectively produce a great many such long sequences of characters. In fact, part of this huge output of very long sequences of characters constitutes the messages actually sent by teletypewriter.

We will, thus, think of all the different Americans who write out telegrams in English as being, approximately at least, an ergodic source of telegraph messages and of all Americans speaking over telephones as being, approximately at least, an ergodic source of telephone signals. Clearly, however, all men writing French plus all men writing English could not constitute an ergodic source. The output of each would have certain time-average probabilities for letters, digrams, trigrams, words, and so on, but the probabilities for the English text would be different from the probabilities for the French text, and the ensemble average would resemble neither.

We will not assert that all writers of English (and all speakers of English) constitute a strictly ergodic message source. The statistics of the English we produce change somewhat as we change subject or purpose, and different people write somewhat differently.

Too, in producing telephone signals by speaking, some people speak softly, some bellow, and some bellow only when they are angry. What we do assert is that we find a remarkable uniformity in many statistics of messages, as in the case of the probability of E for different samples of English text. Speech and writing as ergodic sources are not quite true to the real world, but they are far truer than is the economic man. They are true enough to be useful.

This difference between the exactly ergodic source of the mathematical theory of communication and the approximately ergodic message sources of the real world should be kept in mind. We must exercise a reasonable caution in applying the conclusions of the mathematical theory of communication to actual problems. We are used to this in other fields. For instance, mathematics tells us that we can deduce the diameter of a circle from the coordinates or locations of any three points on the circle, and this is true for absolutely exact coordinates. Yet no sensible man would try to determine the diameter of a somewhat fuzzy real circle drawn on a sheet of paper by trying to measure very exactly the positions of three points a thousandth of an inch apart on its circumference. Rather, he would draw a line through the center and measure the diameter directly as the distance between diametrically opposite points. This is just the sort of judgment and caution one must always use in applying an exact mathematical theory to an inexact practical case.

Whatever caution we invoke, the fact that we have used a random, probabilistic, stochastic process as a model of man in his role of a message source raises philosophical questions. Does this mean that we imply that man acts at random? There is no such implication. Perhaps if we knew enough about a man, his environment, and his history, we could always predict just what word he would write or speak next.

In communication theory, however, we assume that our only knowledge of the message source is obtained either from the messages that the source produces or perhaps from some less-than-complete study of man himself. On the basis of information so obtained, we can derive certain statistical data which, as we have seen, help to narrow the probability as to what the next word or letter of a message will be. There remains an element of uncertainty. For us who have incomplete knowledge of it, the message source behaves *as if* certain choices were made at random, insofar as we cannot predict what the choices will be. If we could predict them, we should incorporate the knowledge which enables us to make the predictions into our statistics of the source. If we had more knowledge, however, we might see that the choices which *we* cannot predict are not really random, in that they are (on the basis of knowledge that we do not have) predictable.

We can see that the view we have taken of finite-state machines, such as that of Figure III-1, has been limited. Finite-state machines can have inputs as well as outputs. The transition from a particular state to one among several others need not be chosen randomly; it could be determined or influenced by various inputs to the machine. For instance, the operation of an electronic digital computer, which is a finite-state machine, is determined by the program and data fed to it by the programmer.

It is, in fact, natural to think that man may be a finite-state machine, not only in his function as a message source which produces words, but in all his other behavior as well. We can think if we like of all possible conditions and configurations of the cells of the nervous system as constituting states (states of mind, perhaps). We can think of one state passing to another, sometimes with the production of a letter, word, sound, or a part thereof, and sometimes with the production of some other action or of some part of an action. We can think of sight, hearing, touch, and other senses as supplying inputs which determine or influence what state the machine passes into next. If man is a finite-state machine, the number of states must be fantastic and beyond any detailed mathematical treatment. But, so are the configurations of the molecules in a gas, and yet we can explain much of the significant behavior of a gas in terms of pressure and temperature merely.

Can we someday say valid, simple, and important things about the working of the mind in producing written text and other things as well? As we have seen, we can already predict a good deal concerning the statistical nature of what a man will write down on paper, unless he is deliberately trying to behave eccentrically, and, even then, he cannot help conforming to habits of his own.

Such broad considerations are not, of course, the real purpose or meat of this chapter. We set out to find a mathematical model adequate to represent some aspects of the human being in his role as a source of messages and adequate to represent some aspects of the messages he produces. Taking English text as an example, we noted that the frequencies of occurrence of various letters are remarkably constant, unless the writer deliberately avoids certain letters. Likewise, frequencies of occurrence of particular pairs, triplets, and so on, of letters are very nearly constant, as are frequencies of various words.

We also saw that we could generate sequences of letters with frequencies corresponding to those of English text by various random or stochastic processes, such as, cutting a lot of text into letters (or words), scrambling the bits of paper in a hat, and drawing them out one at a time. More elaborate stochastic processes, including finite-state machines, can produce an even closer approximation to English text.

Thus, we take a generalized stochastic process as a model of a message source, such as, a source producing English text. But, how must we mathematically define or limit the stochastic sources we deal with so that we can prove theorems concerning the encoding of messages generated by the sources? Of course, we must choose a definition consistent with the character of real English text.

The sort of stochastic source chosen as a model of actual message sources is the ergodic source. An ergodic source can be regarded as a hypothetical machine which produces an infinite number of or ensemble of infinite sequences of characters. Roughly, the nature or statistics of the sequences of characters or messages produced by an ergodic source do not change with time; that is, the source is stationary. Further, for an ergodic source the statistics based on one message apply equally well to all messages that the source generates.

The theorems of communication theory are proved exactly for truly ergodic sources. All writers writing English text together constitute an *approximately* ergodic source of text. The mathematical model—the truly ergodic source—is close enough to the actual situation so that the mathematics we base on it is very useful. But we must be wise and careful in applying the theorems and results of communication theory, which are exact for a mathematical ergodic source, to actual communication problems.

CHAPTER IV Encoding and Binary Digits

A SOURCE OF INFORMATION may be English text, a man speaking, the sound of an orchestra, photographs, motion picture films, or scenes at which a television camera may be pointed. We have seen that in information theory such sources are regarded as having the properties of ergodic sources of letters, numbers, characters, or electrical signals. A chief aim of information theory is to study how such sequences of characters and such signals can be most effectively encoded for transmission, commonly by electrical means.

Everyone has heard of codes and the encoding of messages. Romantic spies use secret codes. Edgar Allan Poe popularized cryptography in *The Gold Bug*. The country is full of amateur cryptanalysts who delight in trying to read encoded messages that others have devised.

In this historical sense of cryptography or secret writing, codes are used to conceal the content of an important message from these for whom it is not intended. This may be done by substituting for the words of the message other words which are listed in a code book. Or, in a type of code called a cipher, letters or numbers may be substituted for the letters in the message according to some previously agreed upon secret scheme.

The idea of encoding, of the accurate representation of one thing by another, occurs in other contexts as well. Geneticists believe that the whole plan for a human body is written out in the chromosomes of the germ cell. Some assert that the "text" consists of an orderly linear arrangement of four different units, or "bases," in the DNA (desoxyribonucleic acid) forming the chromosome. This text in turn produces an equivalent text in RNA (ribonucleic acid), and by means of this RNA text proteins made up of sequences of twenty amino acids are synthesized. Some cryptanalytic effort has been spent in an effort to determine how the four-character message of RNA is reencoded into the twenty-character code of the protein.

Actually, geneticists have been led to such considerations by the existence of information theory. The study of the transmission of information has brought about a new general understanding of the problems of encoding, an understanding which is important to any sort of encoding, whether it be the encoding of cryptography or the encoding of genetic information.

We have already noted in Chapter II that English text can be encoded into the symbols of Morse code and represented by short and long pulses of current separated by short and long spaces. This is one simple form of encoding. From the point of view of information theory, the electromagnetic waves which travel from an FM transmitter to the receiver in your home are an encoding of the music which is transmitted. The electric currents in telephone circuits are an encoding of speech. And the sound waves of speech are themselves an encoding of the motions of the vocal tract which produce them.

Nature has specified the encoding of the motions of the vocal tract into the sounds of speech. The communication engineer, however, can choose the form of encoding by means of which he will represent the sounds of speech by electric currents, just as he can choose the code of dots, dashes, and spaces by means of which he represents the letters of English text in telegraphy. He wants to perform this encoding well, not poorly. To do this he must have some standard which distinguishes good encoding from bad encoding, and he must have some insight into means for achieving good encoding. We learned something of these matters in Chapter II.

It is the study of this problem, a study that might in itself seem limited, which has provided through information theory new ideas important to all encoding, whether cryptographic or genetic. These new ideas include a measure of amount of information, called *entropy,* and a unit of measurement, called the *bit.*

I would like to believe that at this point the reader is clamoring to know the meaning of "amount of information" as measured in bits, and if so I hope that this enthusiasm will carry him over a considerable amount of intervening material about the encoding of messages.

It seems to me that one can't understand and appreciate the solution to a problem unless he has some idea of what the problem is. You can't explain music meaningfully to a man who has never heard any. A story about your neighbor may be full of insight, but it would be wasted on a Hottentot. I think it is only by considering in some detail how a message can be encoded for transmission that we can come to appreciate the need for and the meaning of a measure of amount of information.

It is easiest to gain some understanding of the important problems of coding by considering simple and concrete examples. Of course, in doing this we want to learn something of broad value, and here we may foresee a difficulty.

Some important messages consist of sequences of discrete characters, such as the successive letters of English text or the successive digits of the output of an electronic computer. We have seen, however, that other messages seem inherently different.

Speech and music are variations with time of the pressure of air at the ear. This pressure we can accurately represent in telephony by the voltage of a signal traveling along a wire or by some other quantity. Such a variation of a signal with time is illustrated in *a* of Figure IV-1. Here we assume the signal to be a voltage which varies with time, as shown by the wavy line.

Information theory would be of limited value if it were not applicable to such *continuous* signals or messages as well as to discrete messages, such as English text.

In dealing with continuous signals, information theory first invokes a mathematical theorem called the *sampling theorem,* which we will use but not prove. This theorem states that a continuous signal can be represented completely by and reconstructed perfectly from a set of measurements or *samples* of its amplitude which are made at equally spaced times. The interval between such

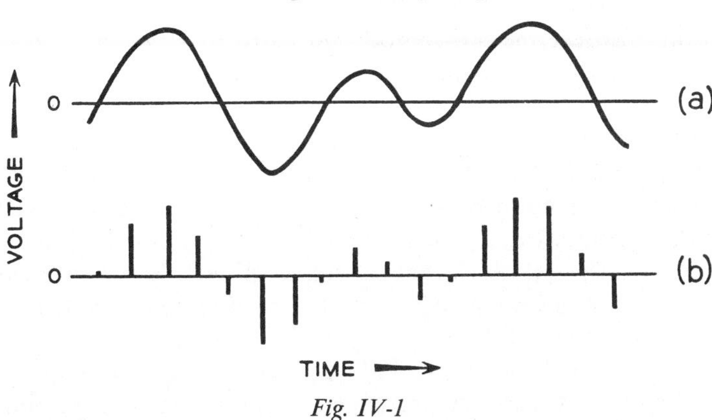

Fig. IV-1

samples must be equal to or less than one-half of the period of the highest frequency present in the signal. A set of such measurements or samples of the amplitude of the signal *a,* Figure IV-1, is represented by a sequence of vertical lines of various heights in *b* of Figure IV-1.

We should particularly note that for such samples of the signal to represent a signal perfectly they must be taken frequently enough. For a voice signal including frequencies from 0 to 4,000 cycles per second we must use 8,000 samples per second. For a television signal including frequencies from 0 to 4 million cycles per second we must use 8 million samples per second. In general, if the frequency range of the signal is f cycles per second we must use at least $2f$ samples per second in order to describe it perfectly.

Thus, the sampling theorem enables us to represent a smoothly varying signal by a sequence of samples which have different amplitudes one from another. This sequence of samples is, however, still inherently different from a sequence of letters or digits. There are only ten digits and there are only twenty-six letters, but a sample can have any of an infinite number of amplitudes. The amplitude of a sample can lie anywhere in a *continuous* range of values, while a character or a digit has only a limited number of *discrete* values.

The manner in which information theory copes with samples having a continuous range of amplitudes is a topic all in itself, to which we will return later. Here we will merely note that a signal need not be described or reproduced perfectly. Indeed, with real physical apparatus a signal *cannot* be reproduced perfectly. In the transmission of speech, for instance, it is sufficient to represent the amplitude of a sample to an accuracy of about 1 per cent. Thus, we can, if we wish, restrict ourselves to the numbers 0 to 99 in describing the amplitudes of successive speech samples and represent the amplitude of a given sample by that one of these hundred integers which is closest to the actual amplitude. By so *quantizing* the signal samples, we achieve a representation comparable to the discrete case of English text.

We can, then, by sampling and quantizing, convert the problem of coding a continuous signal, such as speech, into the seemingly simpler problem of coding a sequence of discrete characters, such as the letters of English text.

We noted in Chapter II that English text can be sent, letter by letter, by means of the Morse code. In a similar manner, such messages can be sent by teletypewriter. Pressing a particular key on the transmitting machine sends a particular sequence of electrical pulses and spaces out on the circuit. When these pulses and spaces reach the receiving machine, they activate the corresponding type bar, and the machine prints out the character that was transmitted.

Patterns of pulses and spaces indeed form a particularly useful and general way of describing or encoding messages. Although Morse code and teletypewriter codes make use of pulses and spaces of different lengths, it is possible to transmit messages by means of a sequence of pulses and spaces of equal length, transmitted at perfectly regular intervals. Figure IV-2 shows how the electric current sent out on the line varies with time for two different patterns, each six intervals long, of such equal pulses and spaces. Sequence *a* is a pulse-space-space-pulse-space-pulse. Sequence *b* is pulse-pulse-pulse-space-pulse-pulse.

The presence of a pulse or a space in a given interval specifies one of two different possibilities. We could use any pair of symbols to represent such patterns of pulses or spaces as those of Figure IV-2: yes, no; +, −; 1, 0. Thus we could represent pattern *a* as follows:

pulse	*space*	*space*	*pulse*	*space*	*pulse*
Yes	*No*	*No*	*Yes*	*No*	*Yes*
+	–	–	+	–	+
1	0	0	1	0	1

The representation by 1 or 0 is particularly convenient and important. It can be used to relate patterns of pulses to numbers expressed in the *binary system* of notation.

When we write 315 we mean

$$\begin{aligned} & 3 \times 10^2 + 1 \times 10^1 + 5 \times 1 \\ = {} & 3 \times 100 + 1 \times 10 + 5 \times 1 \\ = {} & 315 \end{aligned}$$

In this ordinary *decimal* system of representing numbers we make use of the ten different digits: 0, 1, 2, 3, 4, 5, 6, 7, 8, 9. In the binary system we use only two digits, 0 and 1. When we write 1 0 0 1 0 1 we mean

$$\begin{aligned} & 1 \times 2^5 + 0 \times 2^4 + 0 \times 2^3 + 1 \times 2^2 + 0 \times 2 + 1 \times 1 \\ = {} & 1 \times 32 + 0 \times 16 + 0 \times 8 + 1 \times 4 + 0 \times 2 + 1 \times 1 \\ = {} & 37 \text{ in decimal notation} \end{aligned}$$

It is often convenient to let zeros precede a number; this does not change its value. Thus, in decimal notation we can say,

$$0016 = 16$$

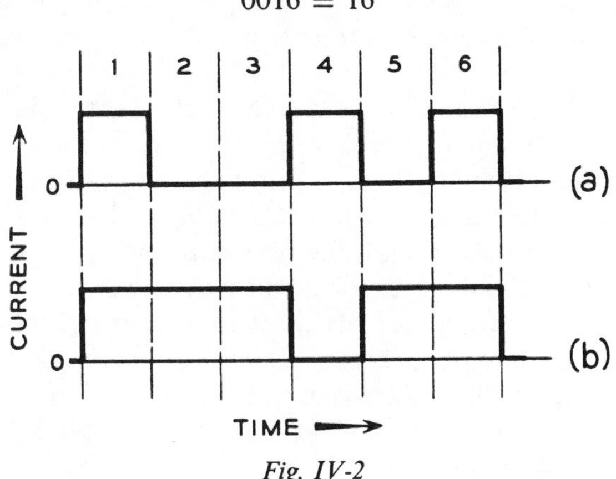

Fig. IV-2

Or in binary notation

$$001010 = 1010$$

In binary numbers, each 0 or 1 is a binary digit. To describe the pulses or spaces occurring in six successive intervals, we can use a sequence of six binary digits. As a pulse or space in one interval is equivalent to a binary digit, we can also refer to a pulse group of six binary digits, or we can refer to the pulse or space occurring in one interval as one binary digit.

Let us consider how many patterns of pulses and spaces there are which are three intervals long. In other words, how many three-digit binary numbers are there? These are all shown in Table VI.

TABLE VI

000	(0)
001	(1)
010	(2)
011	(3)
100	(4)
101	(5)
110	(6)
111	(7)

The decimal numbers corresponding to these sequences of 1's and 0's regarded as binary numbers are shown in parentheses to the right.

We see that there are 8 (0 and 1 through 7) three-digit binary numbers. We may note that 8 is 2^3. We can, in fact, regard an orderly listing of binary digits n intervals long as simply setting down 2^n successive binary numbers, starting with 0. As examples, in Table VII the numbers of different patterns corresponding to different numbers n of binary digits are tabulated.

We see that the number of different patterns increases very rapidly with the number of binary digits. This is because we double the number of possible patterns each time we add one digit. When we add one digit, we get all the old sequences preceded by a 0 plus all the old sequences preceded by a 1.

The binary system of notation is not the only alternative to the

TABLE VII

n (Number of Binary Digits)	*Number of Patterns* (2^n)
1	2
2	4
3	8
4	16
5	32
10	1,024
20	1,048,576

decimal system. The octal system is very important to people who use computers. We can regard the octal system as made up of the eight digits 0, 1, 2, 3, 4, 5, 6, 7.

When we write 356 in the octal system we mean

$$\begin{aligned} & 3 \times 8^2 + 5 \times 8 + 6 \times 1 \\ = & \ 3 \times 64 + 5 \times 8 + 6 \times 1 \\ = & \ 238 \text{ in decimal notation} \end{aligned}$$

We can convert back and forth between the octal and the binary systems very simply. We need merely replace each successive block of three binary digits by the appropriate octal digit, as, for instance,

binary	0 1 0	1 1 1	0 1 1	1 1 0
octal	2	7	3	6

People who work with binary notation in connection with computers find it easier to remember and transcribe a short sequence of octal digits than a long group of binary digits. They learn to regard patterns of three successive binary digits as an entity, so that they will think of a sequence of twelve binary digits as a succession of four patterns of three, that is, as a sequence of four octal digits.

It is interesting to note, too, that, just as a pattern of pulses and spaces can correspond to a sequence of binary digits, so a sequence of pulses of various amplitudes (0, 1, 2, 3, 4, 5, 6, 7) can correspond to a sequence of octal digits. This is illustrated in Figure IV-3. In *a,* we have the sequence of off-on, 0-1 pulses corresponding to the binary number 010111011110. The corresponding octal number is 2736, and in *b* this is represented by a sequence of four pulses of current having amplitudes 2, 7, 3, 6.

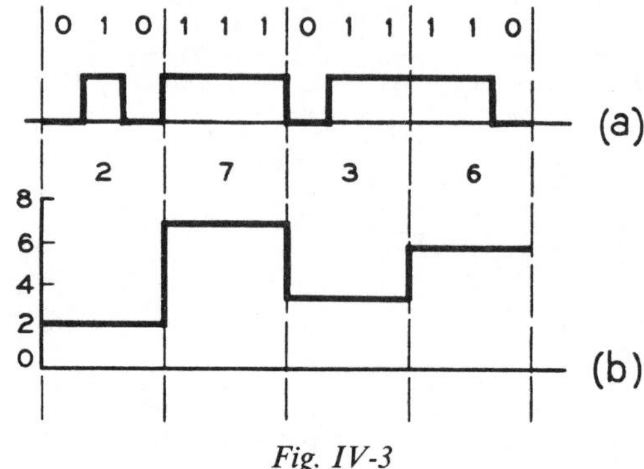

Fig. IV-3

Conversion from binary to decimal numbers is not so easy. On the average, it takes about 3.32 binary digits to represent one decimal digit. Of course we can assign four binary digits to each decimal digit, as shown in Table VIII, but this means that some patterns are wasted; there are more patterns than we use.

It is convenient to think of sequences of 0's and 1's or sequences of pulses and spaces as binary numbers. This helps us to under-

TABLE VIII

Binary Number	*Decimal Digit*
0000	0
0001	1
0010	2
0011	3
0100	4
0101	5
0110	6
0111	7
1000	8
1001	9
1010	not used
1011	not used
1100	not used
1101	not used
1110	not used
1111	not used

stand how many sequences of a different length there are and how numbers written in the binary system correspond to numbers written in the octal or in the decimal system. In the transmission of information, however, the particular number assigned to a sequence of binary digits is irrelevent. For instance, if we wish merely to *transmit* representations of octal digits, we could make the assignments shown in Table IX rather than those in Table VI.

TABLE IX

Sequence of Binary Digits	*Octal Digit Represented*
000	5
001	7
010	1
011	6
100	0
101	4
110	2
111	3

Here the "binary numbers" in the left column designate octal numbers of different numerical value.

In fact, there is another way of looking at such a correspondence between binary digits and other symbols, such as octal digits, a way in which we do not regard the sequence of binary digits as part of a binary number but rather as means of choosing or designating a particular symbol.

We can regard each 0 or 1 as expressing an elementary choice between two possibilities. Consider, for instance, the "tree of choice" shown in Figure IV-4. As we proceed upward from the root to the twigs, let 0 signify that we take the left branch and let 1 signify that we take the right branch. Then 0 1 1 means left, right, right and takes us to the octal digit 6, just as in Table IX.

Just as three binary digits give us enough information to determine one among eight alternatives, four binary digits can determine one among sixteen alternatives, and twenty binary digits can determine one among 1,048,576 alternatives. We can do this by assigning the required binary numbers to the alternatives in any order we wish.

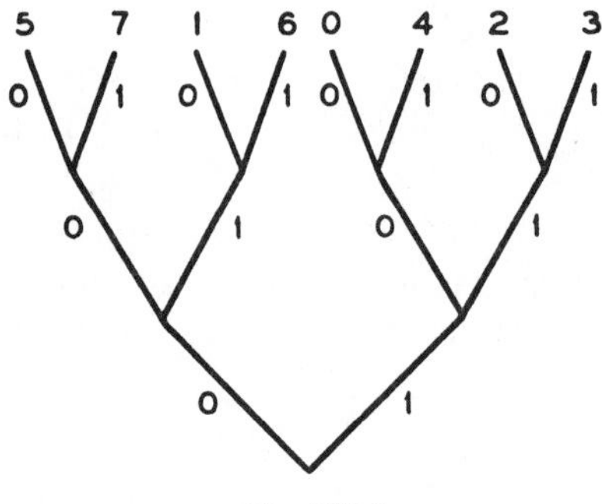

Fig. IV-4

The alternatives which we wish to specify by successions of binary digits need not of course be numbers at all. In fact, we began by considering how we might encode English text so as to transmit it electrically by sequences of pulses and spaces, which can be represented by sequences of binary digits.

A bare essential in transmitting English text letter by letter is twenty-six letters plus a space, or twenty-seven symbols in all. This of course allows us no punctuation and no Arabic numbers.

We can write out the numbers (three, not 3) if we wish and use words for punctuation, (stop, comma, colon, etc.).

Mathematics says that a choice among 27 symbols corresponds to about 4.75 binary digits. If we are not too concerned with efficiency, we can assign a different 5-digit binary number to each character, which will leave five 5-digit binary numbers unused.

My typewriter has 48 keys, including shift and shift lock. We might add two more "symbols" representing carriage return and line advance, making a total of 50. I could encode my actions in typing, capitalization, punctuation, and all (but not insertion of the paper) by a succession of choices among 50 symbols, each choice corresponding to about 5.62 binary digits. We could use 6 binary digits per character and waste some sequences of binary digits.

This waste arises because there are only thirty-two 5-digit binary numbers, which is too few, while there are sixty-four 6-digit binary numbers, which is too many. How can we avoid this waste? If we have 50 characters, we have 125,000 possible different groups of 3 ordered characters. There are 131,072 different combinations of

17 binary digits. Thus, if we divide our text into *blocks* of 3 successive characters, we can specify any possible block by a 17-digit binary number and have a few left over. If we had represented each separate character by 6 binary digits, we would have needed 18 binary digits to represent 3 successive characters. Thus, by this *block coding,* we have cut down the number of binary digits we use in encoding a given length of text by a factor 17/18.

Of course, we might encode English text in quite a different way. We can say a good deal with 16,384 English words. That's quite a large vocabulary. There are just 16,384 fourteen-digit binary numbers. We might assign 16,357 of these to different useful words and 27 to the letters of the alphabet and the space, so that we could spell out any word or sequence of words we failed to include in our word vocabulary. We won't need to put a space between words to which numbers have been assigned; it can be assumed that a space goes with each word.

If we have to spell out words very infrequently, we will use about 14 binary digits per word in this sort of encoding. In ordinary English text there are on the average about 4.5 letters per word. As we must separate words by a space, when we send the message character by character, even if we disregard capitalization and punctuation, we will require on the average 5.5 characters per word. If we encode these using 5 binary digits per character, we will use on the average 27.5 binary digits per word, while in encoding the message word by word we need only 14 binary digits per word.

How can this be so? It is because, in spelling out the message letter by letter, we have provided means for sending with equal facility all sequences of English letters, while, in sending word by word, we restrict ourselves to English words.

Clearly, the average number of binary digits per word required to represent English text depends strongly on how we encode the text.

Now, English text is just one sort of message we might want to transmit. Other messages might be strings of numbers, the human voice, a motion picture, or a photograph. If there are efficient and inefficient ways of encoding English text, we may expect that there will be efficient and inefficient ways of encoding other signals as well.

Indeed, we may be led to believe that there exists in principle some *best* way of encoding the signals from a given message source, a way which will on the average require fewer binary digits per character or per unit time than any other way.

If there is such a best way of encoding a signal, then we might use the average number of binary digits required to encode the signal as a measure of the amount of information per character or the amount of information per second of the message source which produced the signal.

This is just what is done in information theory. How it is done and further reasons for so doing will be considered in the next chapter.

Let us first, however, review very briefly what we have covered in this chapter. In communication theory, we regard coding very broadly, as representing one signal by another. Thus a radio wave can represent the sounds of speech and so form an encoding of these sounds. Encoding is, however, most simply explained and explored in the case of discrete message sources, which produce messages consisting of sequences of characters or numbers. Fortunately, we can represent a continuous signal, such as the current in a telephone line, by a number of samples of its amplitude, using, each second, twice as many samples as the highest frequency present in the signal. Further we can if we wish represent the amplitude of each of these samples approximately by a whole number.

The representation of letters or numbers by sequences of off-or-on signals, which can in turn be represented directly by sequences of the binary digits 0 and 1, is of particular interest in communication theory. For instance, by using sequences of 4 binary digits we can form 16 binary numbers, and we can use 10 of these to represent the 10 decimal digits. Or, by using sequences of 5 binary digits we can form 32 binary numbers, and we can use 27 of these to represent the letters of the English alphabet plus the space. Thus, we can transmit decimal numbers or English text by sending sequences of off-or-on signals.

We should note that while it may be convenient to regard the sequences of binary digits so used as binary numbers, the numerical value of the binary number has no particular significance; we can choose any binary number to represent a particular decimal digit.

If we use 10 of the 16 possible 5-digit binary numbers to encode the 10 decimal digits, we never use (we waste) 6 binary numbers. We could, but never do, transmit these sequences as sequences of off-or-on signals. We can avoid such waste by means of block coding, in which we encode sequences of 2, 3, or more decimal digits or other characters by means of binary digits. For instance, all sequences of 3 decimal digits can be represented by 10 binary digits, while it takes a total of 12 binary digits to represent separately each of 3 decimal digits.

Any sequence of decimal digits may occur, but only certain sequences of English letters ever occur, that is, the words of the English language. Thus, it is more efficient to encode English words as sequences of binary digits rather than to encode the letters of the words individually. This again emphasizes the gain to be made by encoding sequences of characters, rather than encoding each character separately.

All of this leads us to the idea that there may be a best way of encoding the messages from a message source, a way which calls for the least number of binary digits.

CHAPTER V *Entropy*

IN THE LAST CHAPTER, we have considered various ways in which messages can be encoded for transmission. Indeed, all communication involves some sort of encoding of messages. In the electrical case, letters may be encoded in terms of dots or dashes of electric current or in terms of several different strengths of current and directions of current flow, as in Edison's quadruplex telegraph. Or we can encode a message in the binary language of zeros and ones and transmit it electrically as a sequence of pulses or absences of pulses.

Indeed, we have shown that by periodically sampling a continuous signal such as a speech wave and by representing the amplitudes of each sample approximately by the nearest of a set of discrete values, we can represent or encode even such a continuous wave as a sequence of binary digits.

We have also seen that the number of digits required in encoding a given message depends on how it is encoded. Thus, it takes fewer binary digits per character when we encode a group or block of English letters than when we encode the letters one at a time. More important, because only a few combinations of letters form words, it takes considerably fewer digits to encode English text word by word than it does to encode the same text letter by letter.

Surely, there are still other ways of encoding the messages produced by a particular ergodic source, such as a source of English text. How many binary digits per letter or per word are *really* needed? Must we try all possible sorts of encoding in order to find out? But, if we did try all forms of encoding we could think of, we would still not be sure we had found the best form of encoding, for the best form might be one which had not occurred to us.

Is there not, in principle at least, some statistical measurement we can make on the messages produced by the source, a measure which will tell us the minimum average number of binary digits per symbol which will serve to encode the messages produced by the source?

In considering this matter, let us return to the model of a message source which we discussed in Chapter III. There we regarded the message source as an ergodic source of symbols, such as letters or words. Such an ergodic source has certain unvarying statistical properties: the relative frequencies of symbols; the probability that one symbol will follow a particular other symbol, or pair of symbols, or triplet of symbols; and so on.

In the case of English text, we can speak in the same terms of the relative frequencies of words and of the probability that one word will follow a particular word or a particular pair, triplet, or other combination of words.

In illustrating the statistical properties of sequences of letters or words, we showed how material resembling English text can be produced by a sequence of random choices among letters and words, provided that the letters or words are chosen with due regard for their probabilities or their probabilities of following a preceding sequence of letters or words. In these examples, the throw of a die or the picking of a letter out of a hat can serve to "choose" the next symbol.

In writing or speaking, we exercise a similar choice as to what we shall set down or say next. Sometimes we have no choice; Q must be followed by U. We have more choice as to the next symbol in beginning a word than in the middle of a word. However, in any message source, living or mechanical, choice is continually exercised. Otherwise, the messages produced by the source would be predetermined and completely predictable.

Corresponding to the choice exercised by the message source in producing the message, there is an uncertainty on the part of the recipient of the message. This uncertainty is resolved when the recipient examines the message. It is this resolution of uncertainty which is the aim and outcome of communication.

If the message source involved no choice, if, for instance, it could produce only an endless string of ones or an endless string of zeros, the recipient would not need to receive or examine the message to know what it was; he could predict it in advance. Thus, if we are to measure information in a rational way, we must have a measure that increases with the amount of choice of the source and, thus, with the uncertainty of the recipient as to what message the source may produce and transmit.

Certainly, for any message source there are more long messages than there are short messages. For instance, there are 2 possible messages consisting of 1 binary digit, 4 consisting of 2 binary digits, 16 consisting of 4 binary digits, 256 consisting of 8 binary digits, and so on. Should we perhaps say that amount of information should be measured by the number of such messages? Let us consider the case of four telegraph lines used simultaneously in transmitting binary digits between two points, all operating at the same speed. Using the four lines, we can send 4 times as many digits in a given period of time as we could using one line. It also seems reasonable that we should be able to send 4 times as much information by using four lines. If this is so, we should measure information in terms of the number of binary digits rather than in terms of the number of different messages that the binary digits can form. This would mean that amount of information should be measured, not by the number of possible messages, but by the logarithm of this number.

The measure of amount of information which communication theory provides does this and is reasonable in other ways as well. This measure of amount of information is called *entropy.* If we want to understand this entropy of communication theory, it is best first to clear our minds of any ideas associated with the entropy of physics. Once we understand entropy as it is used in communication theory thoroughly, there is no harm in trying to relate it to the entropy of physics, but the literature indicates that some workers have never recovered from the confusion engendered by an early admixture of ideas concerning the entropies of physics and communication theory.

The entropy of communication theory is measured in *bits.* We may say that the entropy of a message source is so many bits per letter, or per word, or per message. If the source produces symbols at a constant rate, we can say that the source has an entropy of so many bits per second.

Entropy increases as the number of messages among which the source may choose increases. It also increases as the freedom of choice (or the uncertainty to the recipient) increases and decreases as the freedom of choice and the uncertainty are restricted. For instance, a restriction that certain messages must be sent either very frequently or very infrequently decreases choice at the source and uncertainty for the recipient, and thus such a restriction must decrease entropy.

It is best to illustrate entropy first in a simple case. The mathematical theory of communication treats the message source as an ergodic process, a process which produces a string of symbols that are to a degree unpredictable. We must imagine the message source as selecting a given message by some random, i.e., unpredictable means, which, however, must be ergodic. Perhaps the simplest case we can imagine is that in which there are only two possible symbols, say, X and Y, between which the message source chooses repeatedly, each choice uninfluenced by any previous choices. In this case we can know only that X will be chosen with some probability p_0 and Y with some probability p_1, as in the outcomes of the toss of a biased coin. The recipient can determine these probabilities by examining a long string of characters (X's, Y's) produced by the source. The probabilities p_0 and p_1 must not change with time if the source is to be ergodic.

For this simplest of cases, the entropy H of the message source is defined as

$$H = -(p_0 \log p_0 + p_1 \log p_1) \text{ bits per symbol}$$

Thus, the entropy is the negative of the sum of the probability p_0 that X will be chosen (or will be received) times the logarithm of p_0 and the probability p_1 that Y will be chosen (or will be received) times the logarithm of this probability.

Whatever plausible arguments one may give for the use of entropy as defined in this and in more complicated cases, the real and true reason is one that will become apparent only as we proceed, and the justification of this formula for entropy will therefore be deferred. It is, however, well to note again that there are different kinds of logarithms and that, in information theory, we use logarithms to the base 2. Some facts about logarithms to the base 2 are noted in Table X.

TABLE X

Fraction p	*Another Way of Writing p*	*Still Another Way of Writing p*	*Log p*
$\frac{3}{4}$	$\frac{1}{2^{.415}}$	$2^{-.415}$	$-.415$
$\frac{1}{2}$	$\frac{1}{2^{1}}$	2^{-1}	-1
$\frac{3}{8}$	$\frac{1}{2^{1.415}}$	$2^{-1.415}$	-1.415
$\frac{1}{4}$	$\frac{1}{2^{2}}$	2^{-2}	-2
$\frac{1}{8}$	$\frac{1}{2^{3}}$	2^{-3}	-3
$\frac{1}{16}$	$\frac{1}{2^{4}}$	2^{-4}	-4
$\frac{1}{64}$	$\frac{1}{2^{6}}$	2^{-6}	-6
$\frac{1}{256}$	$\frac{1}{2^{8}}$	2^{-8}	-8

The logarithm to the base 2 of a number is the power to which 2 must be raised to give the number.

Let us consider, for instance, a "message source" which consists of the tossing of an honest coin. We can let X represent heads and Y represent tails. The probability p_1 that the coin will turn up heads is ½ and the probability p_0 that the coin will turn up tails is also ½. Accordingly, from our expression for entropy and from Table X we find that

$$H = -(\tfrac{1}{2} \log \tfrac{1}{2} + \tfrac{1}{2} \log \tfrac{1}{2})$$
$$H = -[(\tfrac{1}{2})(-1) + (\tfrac{1}{2})(-1)]$$
$$H = 1 \text{ bit per toss}$$

If the message source is the sequence of heads and tails obtained by tossing a coin, it takes one bit of information to convey whether heads or tails has turned up.

Let us notice, now, that we can represent the outcome of successively tossing a coin by a number of binary digits equal to the number of tosses, letting 1 stand for heads and 0 stand for tails. Hence, in this case at least, the entropy, one bit per toss, and the number of binary digits which can represent the outcome, one binary digit per toss, are equal. In this case at least, the number of binary digits necessary to transmit the message generated by the source (the succession of heads and tails) is equal to the entropy of the source.

Suppose the message source produces a string of 1's and 0's by tossing a coin so weighted that it turns up heads ¾ of the time and tails only ¼ of the time. Then

$$
\begin{aligned}
p_1 &= 3/4 \\
p_0 &= 1/4 \\
H &= -(1/4 \log 1/4 + 3/4 \log 3/4) \\
H &= -[(1/4)(-2) + (3/4)(-.415)] \\
H &= .811 \text{ bit per toss}
\end{aligned}
$$

We feel that, in the case of a coin which turns up heads more often than tails, we know more about the outcome than if heads or tails were equally likely. Further, if we were constrained to choose heads more often than tails we would have less choice than if we could choose either with equal probability. We feel that this must be so, for if the probability for heads were 1 and for tails 0, we would have no choice at all. And, we see that the entropy for the case above is only .811 bit per toss. We feel somehow that we ought to be able to represent the outcome of a sequence of such biased tosses by fewer than one binary digit per toss, but it is not immediately clear how many binary digits we must use.

If we choose heads over tails with probability p_1, the probability p_0 of choosing tails must of course be $1 - p_1$. Thus, if we know p_1 we know p_0 as well. We can compute H for various values of p_1 and plot a graph of H vs. p_1. Such a curve is shown in Figure V-1. H has a maximum value of 1 when p_1 is 0.5 and is 0 when p_1 is 0 or 1, that is, when it is certain that the message source always produces either one symbol or the other.

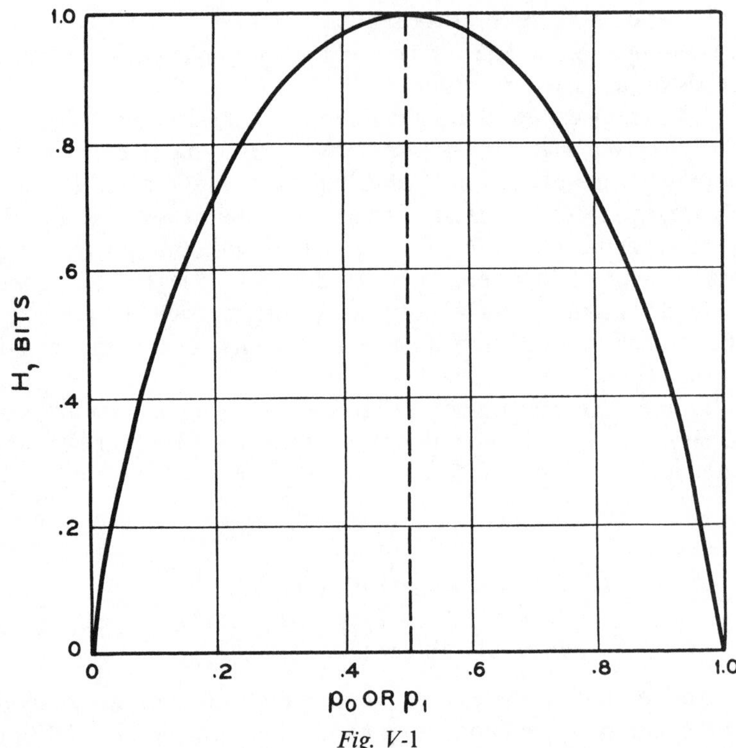

Fig. V-1

Really, whether we call heads X and tails Y or heads Y and tails X is immaterial, so the curve of H vs. p_1 must be the same as H vs. p_0. Thus, the curve of Figure V-1 is symmetrical about the dashed center line at p_1 and p_0 equal to 0.5.

A message source may produce successive choices among the ten decimal digits, or among the twenty-six letters of the alphabet, or among the many thousands of words of the English language. Let us consider the case in which the message source produces one among n symbols or words, with probabilites which are independent of previous choices. In this case the entropy is defined as

$$H = -\sum_{i=1}^{n} p_i \log p_i \text{ bits per symbol} \qquad (5.1)$$

Here the sign Σ (sigma) means to sum or to add up various terms.

p_i is the probability of the ith symbol being chosen. The $i = 1$ below and n above the Σ mean to let i be 1, 2, 3, etc. up to n, so the equation says that the entropy will be given by adding $p_1 \log p_1$ and $p_2 \log p_2$ and so on, including all symbols. We see that when $n = 2$ we have the simple case which we considered earlier.

Let us take an example. Suppose, for instance, that we toss two coins simultaneously. Then there are four possible outcomes, which we can label with the numbers 1 through 4:

H H or 1
H T or 2
T H or 3
T T or 4

If the coins are honest, the probability of each outcome is ¼ and the entropy is

$$H = -(\tfrac{1}{4} \log \tfrac{1}{4} + \tfrac{1}{4} \log \tfrac{1}{4} + \tfrac{1}{4} \log \tfrac{1}{4} + \tfrac{1}{4} \log \tfrac{1}{4})$$
$$H = -(-\tfrac{1}{2} - \tfrac{1}{2} - \tfrac{1}{2} - \tfrac{1}{2})$$
$$H = 2 \text{ bits per pair tossed}$$

It takes 2 bits of information to describe or convey the outcome of tossing a pair of honest coins simultaneously. As in the case of tossing one coin which has equal probabilities of landing heads or tails, we can in this case see that we can use 2 binary digits to describe the outcome of a toss: we can use 1 binary digit for each coin. Thus, in this case too, we can transmit the message generated by the process (of tossing two coins) by using a number of binary digits equal to the entropy.

If we have some number n of symbols all of which are equally probable, the probability of any particular one turning up is $1/n$, so we have n terms, each of which is $1/n \log 1/n$. Thus, the entropy is in this case

$$H = -\log 1/n \quad \text{bits per symbol}$$

For instance, an honest die when rolled has equal probabilities of turning up any number from 1 to 6. Hence, the entropy of the sequence of numbers so produced must be $-\log \tfrac{1}{6}$, or 2.58 bits per throw.

More generally, suppose that we choose each time with equal

likelihood among all binary numbers with N digits. There are 2^N such numbers, so

$$n = 2^N$$

From Table X we easily see that

$$\log 1/n = \log 2^{-N} = -N$$

Thus, for a source which produces at each choice with equal likelihood some N-digit binary number, the entropy is N bits per number. Here the message produced by the source *is* a binary number which can certainly be represented by binary digits. And, again, the message can be represented by a number of binary digits equal to the entropy of the message, measured in bits. This example illustrates graphically how the logarithm *must* be the correct mathematical function in the entropy.

Ordinarily the probability that the message source will produce a particular symbol is different for different symbols. Let us take as an example a message source which produces English words independently of what has gone before but with the probabilities characteristic of English prose. This corresponds to the first-order word approximation given in Chapter III.

In the case of English prose, we find as an empirical fact that if we order the words according to frequency of usage, so that the most frequently used, the most probable word (*the,* in fact) is word number 1, the next most probable word (*of*) is number 2, and so on, then the probability for the rth word is very nearly (if r is not too large)

$$p_r = .1/r \tag{5.2}$$

If equation 5.2 were strictly true, the points in Figure V-2, in which word probability or frequency p_r is plotted against word order or rank r, would fall on the solid line which extends from upper left to lower right. We see that this is very nearly so. This empirical inverse relation between word probability and word rank is known as Zipf's law. We will discuss Zipf's law in Chapter XII; here, we propose merely to use it.

We can show that this equation (5.2) cannot hold for all words. To see this, let us consider tossing a coin. If the probability of heads

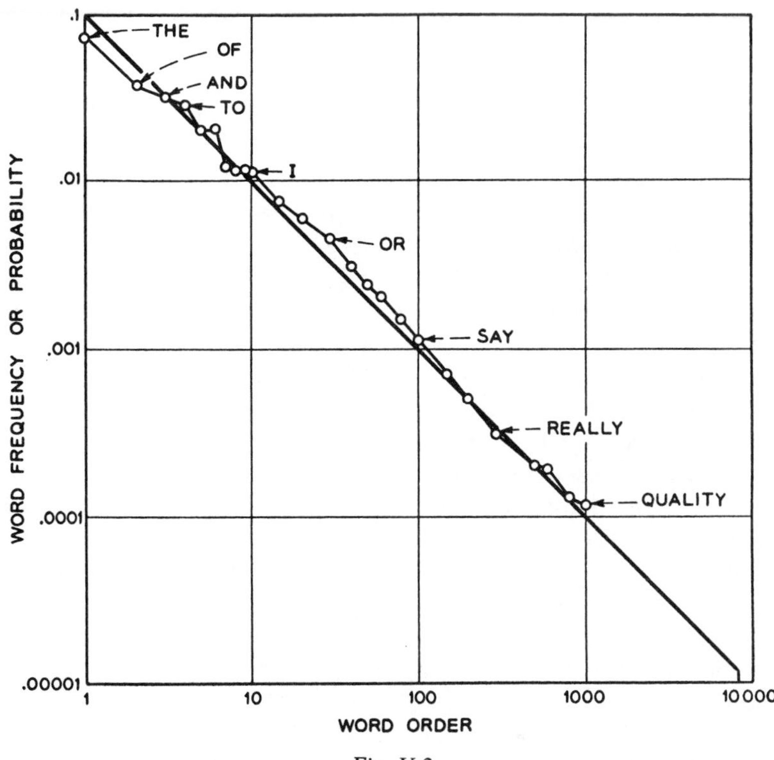

Fig. V-2

turning up is ½ and the probability of tails turning up is ½, then there is no other possible outcome: ½ + ½ = 1. If there were an additional probability of 1/10 that the coin would stand on edge, we would have to conclude that in a hundred tosses we would expect 110 outcomes: heads 50 times, tails 50 times, and standing on edge 10 times. This is patently absurd. The probabilities of all outcomes must add up to unity. Now, let us note that if we add up successively p_1 plus p_2, etc., as given by equation 5.2, we find that by the time we came to p_{8727} the sum of the successive probabilities has become unity. If we took this literally, we would conclude that no additional word could ever occur. Equation 5.1 must be a little in error.

Nonetheless, the error is not great, and Shannon used equation

5.2 in computing the entropy of a message source which produces words independently but with the probability of their occurring in English text. In order to make the sum of the probabilities of all words unity, he included only the 8,727 most frequently used words. He found the entropy to be 9.14 bits per word.

In Chapter IV, we saw that English text can be encoded letter by letter by using 5 binary digits per character or 27.5 binary digits per word. We also saw that by providing different sequences of binary digits for each of 16,357 words and 27 characters, we could encode English text by using about 14 binary digits per word. We are now beginning to suspect that the number of binary digits actually required is given by the entropy, and, as we have seen, Shannon's estimate, based on the relative probabilities of English words, would be 9.14 binary digits per word.

As a next step in exploring this matter of the number of binary digits required to encode the message produced by a message source, we will consider a startling theorem which Shannon proved concerning the "messages" produced by an ergodic source which selects a sequence of letters or words independently with certain probabilities.

Let us consider all of the messages the source can produce which consist of some particular large number of characters. For example, we might consider all messages which are 100,000 symbols (letters, words, characters) long. More generally, let us consider messages having a number M of characters. Some of these messages are more probable than others. In the probable messages, symbol 1 occurs about Mp_1 times, symbol 2 occurs about Mp_2 times, etc. Thus, in these probable messages each symbol occurs with about the frequency characteristic of the source. The source *might* produce other sorts of messages, for instance, a message consisting of one symbol endlessly repeated or merely a message in which the numbers of the various symbols differed markedly from M times their probabilities, but it seldom does.

The remarkable fact is that, if H is the entropy of the source per symbol, there are just about 2^{MH} probable messages, and the rest of the messages all have vanishingly small probabilities of ever occurring. In other words, if we ranked the messages from most probable to least probable, and assigned binary numbers of MH

digits to the 2^{MH} most probable messages, we would be almost certain to have a number corresponding to any M-symbol message that the source actually produced.

Let us illustrate this in particular simple cases. Suppose that the symbols produced are 1 or 0. If these are produced with equal probabilities, a probability ½ that for 1 and a probability ½ that for 0 the entropy H is, as we have seen, 1 bit per symbol. Let us let the source produce messages M digits long. Then $MH = 1{,}000$, and, according to Shannon's theorem, there must be 2^{1000} different probable messages.

Now, by using 1,000 binary digits we can write just 2^{1000} different binary numbers. Thus, in order to assign a different binary number to each probable message, we must use binary numbers 1,000 digits long. This is just what we would expect. In order to designate to the message destination which 1,000 digit binary number the message source produces, we must send a message 1,000 binary digits long.

But, suppose that the digits constituting the messages produced by the message source are obtained by tossing a coin which turns up heads, designating 1, ¾ of the time and tails, designating 0, ¼ of the time. The typical messages so produced will contain more 1's than 0's, but that is not all. We have seen that in this case the entropy H is only .811 bit per toss. If M, the length of the message, is again taken as 1,000 binary digits, MH is only 811. Thus, while as before there are 2^{1000} *possible* messages, there are only 2^{811} *probable* messages.

Now, by using 811 binary digits we can write 2^{811} different binary numbers, and we can assign one of these to each of the 1,000-digit probable messages, leaving the other improbable 1,000-digit messages unnumbered. Thus, we can send word to a message destination which *probable* 1,000-digit message our message source produces by sending only 811 binary digits. And the chance that the message source will produce an improbable 1,000-digit message, to which we have assigned no number, is negligible. Of course, the scheme is not quite foolproof. The message source may still very occasionally turn up a message for which we have no label among all 2^{811} of our 811-digit binary labels. In this case we cannot transmit the message—at least, not by using 811 binary digits.

We see that again we have a strong indication that the number of binary digits required to transmit a message is just the entropy in bits per symbol times the number of symbols. And, we might note that in this last illustration we achieved such an economical transmission by block encoding—that is, by lumping 1,000 (or some other large number) message digits together and representing each probable combination of digits by its individual code (of 811 binary digits).

How firmly and generally can this supposition be established? So far we have considered only cases in which the message source produces each symbol (number, letter, word) independently of the symbols it has produced before. We know this is not true for English text. Besides the constraints of word frequency, there are constraints of word order, so that the writer has less choice as to what the next word will be than he would if he could choose it independently of what has gone before.

How are we to handle this situation? We have a clue in the block coding which we discussed in Chapter IV, and which has been brought to our mind again in the last example. In an ergodic process the probability of the next letter may depend only on the preceding 1, 2, 3, 4, 5, or more letters but not on earlier letters. The second and third order approximations to English given in Chapter III illustrate text produced by such a process. Indeed, in any ergodic process of which we are to make mathematical sense the effect of the past on what symbol will be produced next must decrease as the remoteness of that past is greater. This is reasonably valid in the case of real English as well. While we can imagine examples to the contrary (the consistent use of the same name for a character in a novel), in general the word I write next does not depend on just what word I wrote 10,000 words back.

Now, suppose that before we encode a message we divide it up into very long blocks of symbols. If the blocks are long enough, only the symbols near the beginning will depend on symbols in the previous block, and, if we make the block long enough, these symbols that do depend on symbols in the previous block will form a negligible part of all the symbols in the block. This makes it possible for us to compute the entropy *per block* of symbols by means of equation 5.1. To keep matters straight, let us call the probability of a particular one of the multitudinous long blocks of symbols, which we will call the ith block, $P(B_i)$. Then the entropy per block will be

$$H = -\sum_i P(B_i) \log P(B_i) \quad \text{bits per block}$$

Any mathematician would object to calling this the entropy. He would say, the quantity H given by the above equation *approaches* the entropy as we make the block longer and longer, so that it includes more and more symbols. Thus, we must assume that we make the blocks very long indeed and get a very close approximation to the entropy. With this proviso, we can obtain the entropy per symbol by dividing the entropy per block by the number N of symbols per block

$$H = -(1/N)\sum_i P(B_i) \log P(B_i) \quad \text{bits per symbol} \qquad (5.3)$$

In general, an estimate of entropy is always high if it fails to take into account some relations between symbols. Thus, as we make N, the number of symbols per block, greater and greater, H as given by 5.3 will decrease and approach the true entropy.

We have insisted from the start that amount of information must be so defined that if separate messages are sent over several telegraph wires, the total amount of information must be the sum of the amounts of information sent over the separate wires. Thus, to get the entropy of several message sources operating simultaneously, we add the entropies of the separate sources. We can go further and say that if a source operates intermittently we must multiply its information rate or entropy by the fraction of the time that it operates in order to get its average information rate.

Now, let us say that we have one message source when we have just sent a particular sequence of letters such as TH. In this case the probability that the next letter will be E is very high. We have another particular message source when we have just sent NQ. In this case the probability that the next symbol will be U is unity. We calculate the entropy for each of these message sources. We multiply the entropy of a source which we label B_i by the probability $p(B_i)$ that this source will occur (that is, by the fraction of instances in which this source is in operation). We multiply the entropy of each other source by the probability that that source will occur, and so on. Then we add all the numbers we get in this way in order to get the average entropy or rate of the over-all source, which is a combination of the many different sources, each of which operates only part time. As an example, consider a source involving digram probabilities only, so that the whole effect of the past is summed up in the letter last produced. One source will be the source we have when this letter is E; this will occur in .13 of the total instances. Another source will be the source we have when the letter just produced is W; this will occur in .02 of the total instances.

Putting this in formal mathematical terms, we say that if a particular block of N symbols, which we designate by B_i, has just occurred, the probability that the next symbol will be symbol S_j is

$$p_{B_i}(S_j)$$

The entropy of this "source" which operates only when a particular block of N symbols designated by B_i has just been produced is

$$-\sum_j p_{B_i}(S_j) \log p_{B_i}(S_j)$$

But, in what fraction of instances does this particular message source operate? The fraction of instances in which this source operates is the fraction of instances in which we encounter block B_i rather than some other block of symbols; we call this fraction

$$p(B_i)$$

Thus, taking into account all blocks of N symbols, we write the sum of the entropies of all the separate sources (each separate source defined by what particular block B_i of N symbols has preceded the choice of the symbol S_j) as

$$H_N = -\sum_{i,j} p(B_i)p_{B_i}(S_j) \log p_{B_i}(S_j) \tag{5.4}$$

The i,j under the summation sign mean to let i and j assume all possible values and to add all the numbers we get in this way.

As we let the number N of symbols preceding symbol S_j become very large, H_N approaches the entropy of the source. If there are no statistical influences extending over more than N symbols (this will be true for a digram source for $N = 1$ and for a trigram source for $N = 2$), then H_N is the entropy.

Shannon writes equation 5.4 a little differently. The probability $p(B_i, S_j)$ of encountering the block B_i followed by the symbol S_j is the probability $p(B_i)$ of encountering the block B_i times the probability $p_{B_i}(S_j)$ that symbol S_j will follow block B_i. Hence, we can write 5.4 as follows:

$$H_N = -\sum_{i,j} p(B_i, S_j) \log p_{B_i}(S_j)$$

In Chapter III we consider a finite-state machine, such as that shown in Figure III-3, as a source of text. We can, if we wish, base our computation of entropy on such a machine. In this case, we regard each state of the machine as a message source and compute the entropy for that state. Then we multiply the entropy for that state by the probability that the machine will be in that state and sum (add up) all states in order to get the entropy.

Putting the matter symbolically, suppose that when the machine is in a particular state i it has a probability $p_i(j)$ of producing a particular symbol which we designate by j. For instance, in a state labeled $i = 10$ it might have a probability of 0.03 of producing the third letter of the alphabet, which we label $j = 3$. Then

$$p_{10}(3) = .03$$

The entropy H_i of state i is computed in accord with 5.1:

$$H_i = -\sum_{j} p_i(j) \log p_i(j)$$

Now, we say that the machine has a probability P_i of being in the ith state. The entropy per symbol for the machine as a source of symbols is then

$$H = \sum_{i} P_i H_i \quad \text{bits per symbol}$$

We can write this as

$$H = -\sum_{i,j} P_i p_i(j) \log p_i(j) \quad \text{bits per symbol} \tag{5.5}$$

P_i is the probability that the finite-state machine is in the ith state, and $p_i(j)$ is the probability that it produces the jth symbol when it is in the ith state. The i and j under the Σ mean to allow both i and j to assume all possible values and to add all the numbers so obtained.

Thus, we have gone easily and reasonably from the entropy of a source which produces symbols independently and to which equation 5.1 applies to the more difficult case in which the probability of a symbol occurring depends on what has gone before. And, we have three alternative methods for computing or defining the entropy of the message source. These three methods are equivalent and rigorously correct for true ergodic sources. We should remember, of course, that the source of English text is only approximately ergodic.

Once having defined entropy per symbol in a perfectly general way, the problem is to relate it unequivocally to the average number of binary digits per symbol necessary to encode a message.

We have seen that if we divide the message into a block of letters or words and treat each possible block as a symbol, we can compute the entropy per block by the same formula we used per independent symbol and get as close as we like to the source entropy merely by making the blocks very long.

Thus, the problem is to find out how to encode efficiently in binary digits a sequence of symbols chosen from a very large group of symbols, each of which has a certain probability of being chosen. Shannon and Fano both showed ways of doing this, and Huffman found an even better way, which we shall consider here.

Let us for convenience list all the symbols vertically in order of decreasing probability. Suppose the symbols are the eight words *the, man, to, runs, house, likes, horse, sells,* which occur independently with probabilities of their being chosen, or appearing, as listed in Table XI.

We can compute the entropy per word by means of 5.1; it is 2.21 bits per word. However, if we merely assigned one of the eight 3-digit binary numbers to each word, we would need 3 digits to transmit each word. How can we encode the words more efficiently?

Figure V-3 shows how to construct the most efficient code for encoding such a message *word by word.* The words are listed to the

TABLE XI

Word	*Probability*
the	.50
man	.15
to	.12
runs	.10
house	.04
likes	.04
horse	.03
sells	.02

left, and the probabilities are shown in parentheses. In constructing the code, we first find the two lowest probabilities, .02 (*sells*) and .03 (*horse*), and draw lines to the point marked .05, the probability of either *horse* or *sells.* We then disregard the individual probabilities connected by the lines and look for the two lowest probabilities, which are .04 (like) and .04 (house). We draw lines to the right to a point marked .08, which is the sum of .04 and .04. The two lowest remaining probabilities are now .05 and .08, so we draw a line to the right connecting them, to give a point marked

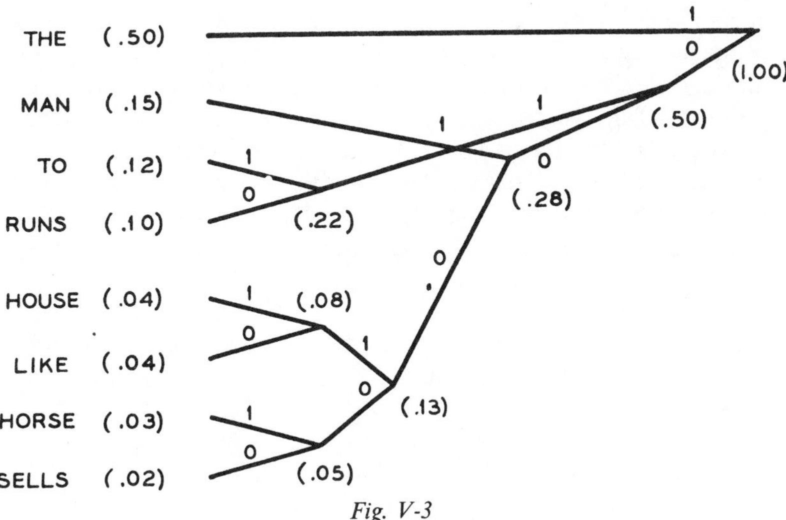

Fig. V-3

.13. We proceed thus until paths run from each word to a common point to the right, the point marked 1.00. We then label each upper path going to the left from a point 1 and each lower path 0. The code for a given word is then the sequence of digits encountered going left from the common point 1.00 to the word in question. The codes are listed in Table XII.

TABLE XII

Word	*Probability p*	*Code*	*Number of Digits in Code, N*	*Np*
the	.50	1	1	.50
man	.15	001	3	.45
to	.12	011	3	.36
runs	.10	010	3	.30
house	.04	00011	5	.20
likes	.04	00010	5	.20
horse	.03	00001	5	.15
sells	.02	00000	5	.10
				2.26

In Table XII we have shown not only each word and its code but also the probability of each code and the number of digits in each code. The probability of a word times the number of digits in the code gives the average number of digits per word in a long message due to the use of that particular word. If we add the products of the probabilities and the numbers of digits for all the words, we get the average number of digits per word, which is 2.26. This is a little larger than the entropy per word, which we found to be 2.21 bits per word, but it is a smaller number of digits than the 3 digits per word we would have used if we had merely assigned a different 3-digit code to each word.

Not only can it be proved that this Huffman code is the most efficient code for encoding a set of symbols having different probabilities, it can be proved that it always calls for less than one binary digit per symbol more than the entropy (in the above example, it calls for only 0.05 extra binary digits per symbol).

Now suppose that we combine our symbols into blocks of 1, 2, 3, or more symbols before encoding. Each of these blocks will have a probability (in the case of symbols chosen independently, the probability of a sequence of symbols will be the product of the probabilities of the symbols). We can find a Huffman code for these blocks of symbols. As we make the blocks longer and longer, the number of binary digits in the code for each block will increase. Yet, our Huffman code will take less than one extra digit per block above the entropy in bits per block! Thus, as the blocks and their codes become very long, the less-than-one extra digit of the Huffman code will become a negligible fraction of the total number of digits, and, as closely as we like (by making the blocks longer), the number of binary digits per block will equal the entropy in bits per block.

Suppose we have a communication channel which can transmit a number *C* of off-or-on pulses per second. Such a channel can transmit *C* binary digits per second. Each binary digit is capable of transmitting one bit of information. Hence we can say that the information *capacity* of this communication channel is *C* bits per second. If the entropy *H* of a message source, measured in bits per second, is less than *C*, then, by encoding with a Huffman code, the signals from the source can be transmitted over the channel.

Not all channels transmit binary digits. A channel, for instance, might allow three amplitudes of pulses, or it might transmit different pulses of different lengths, as in Morse code. We can imagine connecting various different message sources to such a channel. Each source will have some entropy or information rate. Some source will give the highest entropy that can be transmitted over the channel, and this highest possible entropy is called the *channel capacity C* of the channel and is measured in bits per second.

By means of the Huffman code, the output of the channel when it is transmitting a message of this greatest possible entropy can be coded into some least number of binary digits per second, and, when long stretches of message are encoded into long stretches of binary digits, it must take very close to *C* binary digits per second to represent the signals passing over the channel.

This encoding can, of course, be used in the reverse sense, and *C* independent binary digits per second can be so encoded as to be transmitted over the channel. Thus, a source of entropy *H* can be encoded into *H* binary digits per second, and a general discrete channel of capacity *C* can be used to transmit *C* bits per second.

We are now in a position to appreciate one of the fundamental theorems of information theory. Shannon calls this the fundamental theorem of the noiseless channel. He states it as follows:

> Let a source have entropy H (bits per symbol) and a channel have a capacity [to transmit] C bits per second. Then it is possible to encode the ousput of the source in such a way as to transmit at the average rate (C/H) − ε symbols per second over the channel, where ε is arbitrarily small. It is not possible to transmit at an average rate greater than C/H.

Let us restate this without mathematical niceties. Any discrete channel that we may specify, whether it transmits binary digits, letters and numbers, or dots, dashes, and spaces of certain distinct lengths has some particular unique channel capacity *C*. Any ergodic message source has some particular entropy *H*. If *H* is less than or equal to *C*, we can transmit the messages generated by the source over the channel. If *H* is greater than *C*, we had better not try to do so, because we just plain can't.

We have indicated above how the first part of this theorem can be proved. We have not shown that a source of entropy *H* cannot be encoded in less than *H* binary digits per symbol, but this also can be proved.

We have now firmly arrived at the fact that the entropy of a message source measured in bits tells us how many binary digits (or off-or-on pulses, or yeses-or-noes) are required, per character, or per letter, or per word, or per second in order to transmit messages produced by the source. This identification goes right back to Shannon's original paper. In fact, the word *bit* is merely a contraction of *binary digit* and is generally used in place of *binary digit.*

Here I have used *bit* in a particular sense, as a measure of amount of information, and in other contexts I have used a different expression, binary digit. I have done this in order to avoid a confusion which might easily have arisen had I started out by using *bit* to mean two different things.

After all, in practical situations the entropy in bits is usually different from the number of binary digits involved. Suppose, for instance, that a message source randomly produces the symbol 1 with a probability ¼ and the symbol 0 with the probability ¾ and that it produces 10 symbols per second. Certainly such a source produces binary digits at a rate of 10 per second, but the information rate or entropy of the source is .811 bit per binary digit and 8.11 bits per second. We could encode the sequence of binary digits produced by this source by using on the average only 8.11 binary digits per second.

Similarly, suppose we have a communication channel which is capable of transmitting 10,000 arbitrarily chosen off-or-on pulses per second. Certainly, such a channel has a channel capacity of 10,000 bits per second. However, if the channel is used to transmit a completely repetitive pattern of pulses, we must say that the actual rate of transmission of information is 0 bits per second, despite the fact that the channel is certainly transmitting 10,000 binary digits per second.

Here we have used bit only in the sense of a binary measure of amount of information, as a measure of the entropy or information rate of a message source in bits per symbol or in bits per second or as a measure of the information transmission capabilities of a channel in bits per symbol or bits per second. We can describe it as an elementary binary choice or decision among two possibilities which have equal probabilities. At the message source a bit represents a certain amount of choice as to the message which will be generated; in writing grammatical English we have on the average a choice of about one bit per letter. At the destination a bit of information resolves a certain amount of uncertainty; in receiving English text there is on the average, about one bit of uncertainty as to what the next letter will be.

When we are transmitting messages generated by an information source by means of off-or-on pulses, we know how many binary digits we are transmitting per second even when (as in most cases) we don't know the entropy of the source. (If we know the entropy of the source in bits per second to be less than the binary digits used per second, we would know that we could get along in principle with fewer binary digits per second.) We know how to use the binary digits to specify or determine one out of several possibilities, either by means of a tree such as that of Figure IV-4 or by means of a Huffman code such as that of Figure V-3. It is common in such a case to speak of the rate of transmission of binary digits as a bit rate, but there is a certain danger that the inexperienced may muddy their thinking if they do this.

All that I really ask of the reader is to remember that we have used *bit* in one sense only, as a measure of information and have called 0 or 1 a binary digit. If we can transmit 1,000 freely chosen binary digits per second, we can transmit 1,000 bits of information a second. It may be convenient to use *bit* to mean *binary digit,* but when we do so we should be sure that we understand what we are doing.

Let us now return for a moment to an entirely different matter, the Huffman code given in Table XII and Figure V-3. When we encode a message by using this code and get an uninterrupted string of symbols, how do we tell whether we should take a particular 1 in the string of symbols as indicating the word *the* or as part of the code for some other word?

We should note that of the codes in Table XII, none forms the first part of another. This is called the *prefix property.* It has important and, indeed, astonishing consequences, which are easily illustrated. Suppose, for instance, that we encode the message: the man sells the house to the man the horse runs to the man. The encoded message is as follows:

|the| man | sells |the| house |
1 0 0 1 0 0 0 0 0 1 0 0 0 1 1
| likes | man |the|

to |the| man |the| horse |
0 1 1 1 0 0 1 1 0 0 0 0 1
to |the| man |the| horse |

runs | to |the| man |
0 1 0 0 1 1 1 0 0 1
runs | to |the| man |

Here the message words are written above the code groups.

Now suppose we receive only the digits following the first vertical dashed line below the digits. We start to decode by looking for the shortest sequence of digits which constitutes a word in our code. This is 00010, which corresponds to *likes.* We go on in this fashion. The "decoded" words are written under the code, separated by dashed lines.

We see that after a few errors the dashed lines correspond to the solid lines, and from that point on the deciphered message is correct. We don't need to know where the message starts in order to decode it as correctly as possible (unless all code words are of equal length).

When we look back we can see that we have fulfilled the purpose of this chapter. We have arrived at a measure of the amount of information per symbol or per unit time of an ergodic source, and we have shown how this is equal to the average number of binary digits per symbol necessary to transmit the messages produced by the source. We have noted that to attain transmission with negligibly more bits than the entropy, we must encode the messages produced by the source in long blocks, not symbol by symbol.

We might ask, however, how long do the blocks have to be? Here we come back to another consideration. There are two reasons for encoding in long blocks. One is, in order to make the average number of binary digits per symbol used in the Huffman code negligibly larger than the entropy per symbol. The other is, that to encode such material as English text efficiently we must take into account the influence of preceding symbols on the probability that a given symbol will appear next. We have seen that we can do this using equation 5.3 and taking very long blocks.

We return, then, to the question: how many symbols N must the block of characters have so that (1) the Huffman code is very efficient, (2) the entropy per block, disregarding interrelations outside of the block, is very close to N times the entropy per symbol? In the case of English text, condition 2 is governing.

Shannon has estimated the entropy per letter for English text by measuring a person's ability to guess the next letter of a message after seeing 1, 2, 3, etc., preceding letters. In these texts the "alphabet" used consisted of 26 letters plus the space.

Figure V-4 shows the upper and lower bounds on the entropy of English plotted vs. the number of letters the person saw in

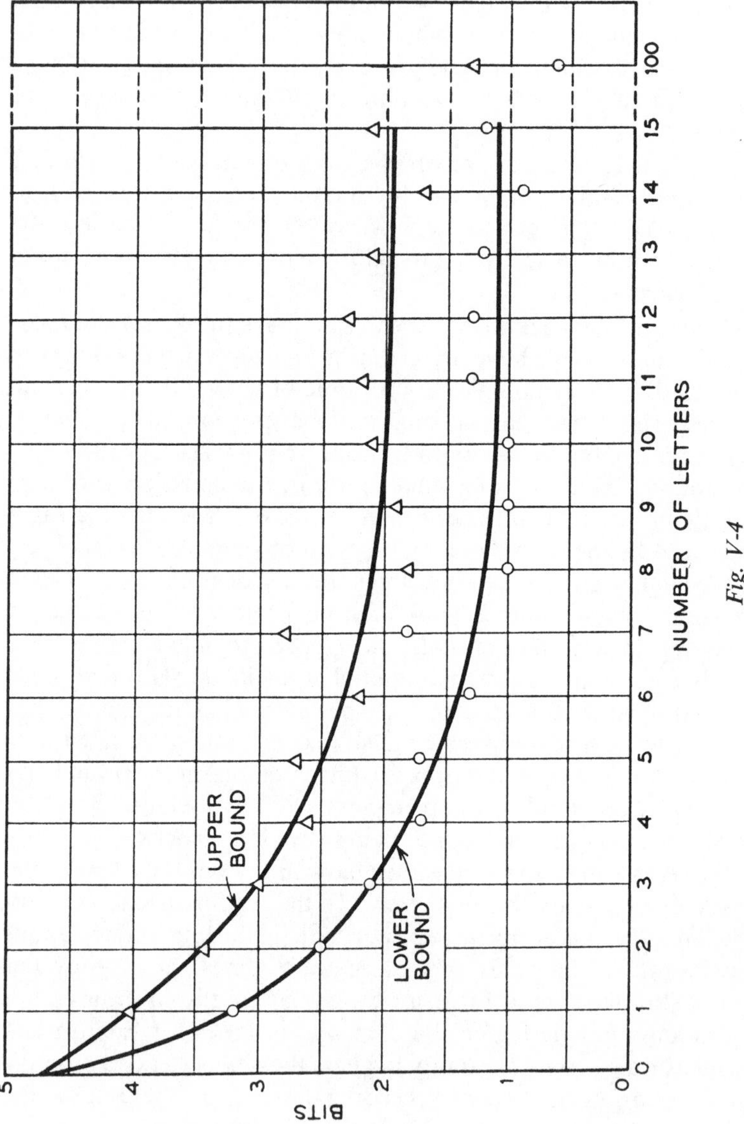
BITS
NUMBER OF LETTERS
UPPER BOUND
LOWER BOUND
0
1
2
3
4
5
6
7
8
9
10
11
12
13
14
15
100

Fig. V-4

making his prediction. While the curve seems to drop slowly as the number of letters is increased from 10 to 15, it drops substantially between 15 and 100. This would appear to indicate that we might have to encode in blocks as large as 100 letters long in order to encode English really efficiently.

From Figure V-4 it appears that the entropy of English text lies somewhere between 0.6 and 1.3 bits per letter. Let us assume a value of 1 bit per letter. Then it will take on the average 100 binary digits to encode a block of 100 letters. This means that there are 2^{100} probable English sequences of 100 letters. In our usual decimal notation, 2^{100} can be written as 1 followed by 30 zeroes, a fantastically large number.

In endeavoring to find the probability in English text of all meaningful blocks of letters 100 letters long, we would have to count the relative frequency of occurrence of each such block. Since there are 10^{30} highly likely blocks, this would be physically impossible.

Further, this is impossible in principle. Most of these 10^{30} sequences of letters and spaces (which do not include *all* meaningful sequences) have never been written down! Thus, it is impossible to speak of their relative frequencies or probabilities of such long blocks of letters as derived from English text.

Here we are really confronted with two questions: the accuracy of the description of English text as the product of an ergodic source and the most appropriate statistical description of that source. One may believe that appropriate probabilities do exist in some form in the human being even if they cannot be evaluated by the examination of existing text. Or one may believe that the probabilities exist and that they can be derived from data taken in some way more appropriate than a naïve computation of the probabilities of sequences of letters. We may note, for instance, that equations 5.4 and 5.5 also give the entropy of an ergodic source. Equation 5.5 applies to a finite-state machine. We have noted at the close of Chapter III that the idea of a human being being in some particular state and in that state producing some particular symbol or word is an appealing one.

Some linguists hold, however, that English grammar is inconsistent with the output of a finite-state machine. Clearly, in trying to understand the structure and the entropy of actual English text we would have to consider such text much more deeply than we have up to this point.

It is safe if not subtle to apply an exact mathematical theory blindly and mechanically to the ideal abstraction for which it holds. We must be clever and wise in using even a good and appropriate mathematical theory in connection with actual, nonideal problems. We should seek a simple and realistic description of the laws governing English text if we are to relate it with communication theory as successfully as possible. Such a description must certainly involve the grammar of the language, which we will discuss in the next chapter.

In any event, we know that there are some valid statistics of English text, such as letter and word frequencies, and the coding theorems enable us to take advantage of such known statistics.

If we encode English letter by letter, disregarding the relative frequencies of the letters, we require 4.76 binary digits per character (including space). If we encode letter by letter, taking into account the relative probabilities of various letters, we require 4.03 binary digits per character. If we encode word by word, taking into account relative frequencies of words, we require 1.66 binary digits per character. And, by using an ingenious and appropriate means, Shannon has estimated the entropy of English text to be between .6 and 1.3 bits per letter, so that we may hope for even more efficient encoding.

If, however, we mechanically push some particular procedure for finding the entropy of English text to the limit, we can easily engender not only difficulties but nonsense. Perhaps we can ascribe this nonsense partly to differences between man as a source of English text and our model of an ideal ergodic source, but partly we should ascribe it to the use of an inappropriate approach. We can surely say that the model of man as an ergodic source of text is good and useful if not perfect, and we should regard it highly for these qualities.

This chapter has been long and heavy going, and a summary seems in order. Clearly, it is impossible to recapitulate briefly all those matters which took so many pages to expound. We can only re-emphasize the most vital points.

In communication theory the entropy of a signal source in bits per symbol or per second gives the average number of binary digits, per symbol or per second, necessary to encode the messages produced by the source.

We think of the message source as randomly, that is, unpredictably, choosing one among many possible messages for transmission. Thus, in connection with the message source we think of entropy as a measure of choice, the amount of choice the source excercises in selecting the one particular message that is actually transmitted.

We think of the recipient of the message, prior to the receipt of the message, as being uncertain as to which among the many possible messages the message source will actually generate and transmit to him. Thus, we think of the entropy of the message source as measuring the uncertainty of the recipient as to which message will be received, an uncertainty which is resolved on receipt of the message.

If the message is one among n equally probable symbols or messages, the entropy is log n. This is perfectly natural, for if we have log n binary digits, we can use them to write out

$$2^{\log n} = n$$

different binary numbers, and one of these numbers can be used as a label for each of the n messages.

More generally, if the symbols are not equally probable, the entropy is given by equation 5.1. By regarding a very long block of symbols, whose content is little dependent on preceding symbols, as a sort of super symbol, equation 5.1 can be modified to give the entropy per symbol for information sources in which the probability that a symbol is chosen depends on what symbols have been chosen previously. This gives us equation 5.3. Other general expressions for entropy are given by equations 5.4 and 5.5.

By assuming that the symbols or blocks of symbols which a source produces are encoded by a most efficient binary code called a Huffman code, it is possible to prove that the entropy of an ergodic source measured in bits is equal to the average number of binary digits necessary to encode it.

An error-free communication channel may not transmit binary

digits; it may transmit letters or other symbols. We can imagine attaching different message sources to such a channel and seeking (usually mathematically) the message source that causes the entropy of the message transmitted over the channel to be as large as possible. This largest possible entropy of a message transmitted over an error-free channel is called the channel capacity. It can be proved that, if the entropy of a source is less than the channel capacity of the channel, messages from the source can be encoded so that they can be transmitted over the channel. This is Shannon's fundamental theorem for the noiseless channel.

In principle, expressions such as equations 5.1, 5.3, 5.4, and 5.5 enable us to compute the entropy of a message source by statistical analysis of messages produced by the source. Even for an ideal ergodic source, this would often call for impractically long computations. In the case of an actual source, such as English text, some naïve prescriptions for computing entropy can be meaningless.

An approximation to the entropy can be obtained by disregarding the effect of some past symbols on the probability of the source producing a particular symbol next. Such an approximation to the entropy is always too large and calls for encoding by means of more binary digits than are absolutely necessary. Thus, if we encode English text letter by letter, disregarding even the relative probabilities of letters, we require 4.76 binary digits per letter, while if we encode word by word, taking into account the relative probability of words, we require 1.66 binary digits per letter.

If we wanted to do even better we would have to take into account other features of English such as the effect of the constraints imposed by grammar on the probability that a message source will produce a particular word.

While we do not know how to encode English text in a highly efficient way, Shannon made an ingenious experiment which shows that the entropy of English text must lie between .6 and 1.3 bits per character. In this experiment a person guessed what letter would follow the letters of a passage of text many letters long.

CHAPTER VI *Language and Meaning*

THE TWO GREAT TRIUMPHS of information theory are establishing the channel capacity and, in particular, the number of binary digits required to transmit information from a particular source and showing that a noisy communication channel has an information rate in bits per character or bits per second up to which errorless transmission is possible despite the noise. In each case, the results must be demonstrated for discrete and for continuous sources and channels.

After four chapters of by no means easy preparation, we were finally ready to essay in the previous chapter the problem of the number of binary digits required to transmit the information generated by a truly ergodic discrete source. Were this book a text on information theory, we would proceed to the next logical step, the noisy discrete channel, and then on to the ergodic continuous channel.

At the end of such a logical progress, however, our thoughts would necessarily be drawn back to a consideration of the message sources of the real world, which are only approximately ergodic, and to the estimation of their entropy and the efficient encoding of the messages they produce.

Rather than proceeding further with the strictly mathematical aspects of communication theory at this point, is it not more attractive to pause and consider that chief form of communication, language, in the light of communication theory? And, in doing so, why should we not let our thoughts stray a little in viewing an important part of our world from the small eminence we have attained? Why should we not see whether even the broad problems of language and meaning seem different to us in the light of what we have learned?

In following such a course the reader should heed a word of caution. So far the main emphasis has been on what we *know*. What we know is the hard core of science. However, scientists find it very difficult to share the things that they know with laymen. To understand the sure and the reasonably sure knowledge of science takes the sort of hard thought which I am afraid was required of the reader in the last few chapters.

There is, however, another and easier though not entirely frivolous side to science. This is a peculiar type of informed ignorance. The scientist's ignorance is rather different from the layman's ignorance, because the background of established fact and theory on which the scientist bases his peculiar brand of ignorance excludes a wide range of nonsense from his speculations. In the higher and hazier reaches of the scientist's ignorance, we have scientifically informed ignorance about the origin of the universe, the ultimate basis of knowledge, and the relation of our present scientific knowledge to politics, free will, and morality. In this particular chapter we will dabble in what I hope to be scientifically informed ignorance about language.

The warning is, of course, that much of what will be put forward here about language is no more than informed ignorance. The warning seems necessary because it is very hard for laymen to tell scientific ignorance from scientific fact. Because the ignorance is necessarily expressed in broader, sketchier, and less qualified terms than is the fact, it is easier to assimilate. Because it deals with grand and unsolved problems, it is more romantic. Generally, it has a wider currency and is held in higher esteem than is scientific fact.

However hazardous such ignorance may be to the layman, it is valuable to the scientist. It is this vision of unattained lands, of unscaled heights, which rescues him from complacency and spurs him beyond mere plodding. But when the scientist is airing his ignorance he usually knows what he is doing, while the unwarned layman apparently often does not and is left scrambling about on cloud mountains without ever having set foot on the continents of knowledge.

With this caution in mind, let us return to what we have already encountered concerning language and proceed thence.

In what follows we will confine ourselves to a discussion of grammatical English. We all know (and especially those who have had the misfortune of listening to a transcription of a seemingly intelligible conversation or technical talk) that much spoken English appears to be agrammatical, as, indeed, much of Gertrude Stein is. So are many conventions and clichés. "Me heap big chief" is perfectly intelligible anywhere in the country, yet it is certainly not grammatical. Purists do not consider the inverted word order which is so characteristic of second-rate poetry as being grammatical.

Thus, a discussion of grammatical English by no means covers the field of spoken and written communication, but it charts a course which we can follow with some sense of order and interest.

We have noted before that, if we are to write what will be accepted as English text, certain constraints must be obeyed. We cannot simply set down any word following any other. A complete grammar of a language would have to express all of these constraints fully. It should allow within its rules the construction of any sequence of English words which will be accepted, at some particular time and according to some particular standard, as grammatical.

The matter of acceptance of constructions as grammatical is a difficult and hazy one. The translators who produced the King James Bible were free to say "fear not," "sin not," and "speak not" as well as "think not," "do not," or "have not," and we frequently repeat the aphorism "want not, waste not." Yet in our everyday speech or writing we would be constrained to say "do not fear," "do not sin," or "do not speak," and we might perhaps say, "If you are not to want, you should not waste." What is grammatical certainly changes with time. Here we can merely notice this and pass on to other matters.

Certainly, a satisfactory grammar must prescribe certain rules which allow the construction of all possible grammatical utterances and of grammatical utterances only. Besides doing this, satisfactory rules of grammar should allow us to analyze a sentence so as to distinguish the features which were determined merely by the rules of grammar from any other features.

If we once had such rules, we would be able to make a new estimate of the entropy of English text, for we could see what part of sentence structure is a mere mechanical following of rules and what part involves choice or uncertainty and hence contributes to entropy. Further, we could transmit English efficiently by transmitting as a message only data concerning the choices exercised in constructing sentences; at the receiver, we could let a grammar machine build grammatical sentences embodying the choices specified by the received message.

Even grammar, of course, is not the whole of language, for a sentence can be very odd even if it is grammatical. We can imagine that, if a machine capable of producing only grammatical sentences made its choices at random, it might perhaps produce such a sentence as "The chartreuse semiquaver skinned the feelings of the manifold." A man presumably makes his choices in some other way if he says, "The blue note flayed the emotions of the multitude." The difference lies in what choices one makes while following grammatical rules, not in the rules themselves. An understanding of grammar would not unlock to us all of the secrets of language, but it would take us a long step forward.

What sort of rules will result in the production of grammatical sentences only and of all grammatical sentences, even when choices are made at random? In Chapter III we saw that English-like sequences of words can be produced by choosing a word at random according to its probability of succeeding a preceding sequence of words some *M* words long. An example of a second-order word approximation, in which a word is chosen on the basis of its succeeding the previous word, was given.

One can construct higher-order word approximations by using the knowledge of English which is stored in our heads. One can, for instance, obtain a fourth-order word approximation by simply showing a sequence of three connected words to a person and asking him to think up a sentence in which the sequence of words occurs and to add the next word. By going from person to person a long string of words can be constructed, for instance:

1. When morning broke after an orgy of wild abandon he said her head shook vertically aligned in a sequence of words signifying what.

2. It happened one frosty look of trees waving gracefully against the wall.

3. When cooked asparagus has a delicious flavor suggesting apples.

4. The last time I saw him when he lived.

These "sentences" are as sensible as they are because selections of words were not made at random but by thinking beings. The point to be noted is how astonishingly grammatical the sentences are, despite the fact that rules of grammar (and sense) were applied to only four words at a time (the three shown to each person and the one he added). Still, example 4 is perhaps dubiously grammatical.

If Shannon is right and there is in English text a choice of about 1 bit per symbol, then choosing among a group of 4 words could involve about 22 binary choices, or a choice among some 10 million 4-word combinations. In principle, a computer could be made to add words by using such a list of combinations, but the result would not be assuredly grammatical, nor could we be sure that this cumbersome procedure would produce all possible grammatical sequences of words. There probably are sequences of words which could form a part of a grammatical sentence in one case and could not in another case. If we included such a sequence, we would produce some nongrammatical sentences, and, if we excluded it, we would fail to produce all grammatical sentences.

If we go to combinations of more than four words, we will favor grammar over completeness. If we go to fewer than four words, we will favor completeness over grammar. We can't have both.

The idea of a finite-state machine recurs at this point. Perhaps at each point in a sentence a sentence-producing machine should be in a particular state, which allows it certain choices as to what state it will go to next. Moreover, perhaps such a machine can deal with certain classes or subclasses of words, such as singular nouns, plural nouns, adjectives, adverbs, verbs of various tense and number, and so on, so as to produce grammatical structures into which words can be fitted rather than sequences of particular words.

The idea of grammar as a finite-state machine is particularly appealing because a mechanist would assert that man must be a finite-state machine, because he consists of only a finite number of cells, or of atoms if we push the matter further.

Noam Chomsky, a brilliant and highly regarded modern linguist, rejects the finite-state machine as either a possible or a proper model of grammatical structure. Chomsky points out that there are many rules for constructing sequences of characters which cannot be embodied in a finite-state machine. For instance, the rule might be, choose letters at random and write them down until the letter Z shows up, then repeat all the letters since the preceding Z in reverse order, and then go on with a new set of letters, and so on. This process will produce a sequence of letters showing clear evidence of long-range order. Further, there is no limit to the possible length of the sequence between Z's. No finite-state machine can simulate this process and this result.

Chomsky points out that there is no limit to the possible length of grammatical sentences in English and argues that English sentences are organized in such a way that this is sufficient to rule out a finite-state machine as a source of all possible English text. But, can we really regard a sentence miles long as grammatical when we know darned well that no one ever has or will produce such a sentence and that no one could understand it if it existed?

To decide such a question, we must have a standard of being grammatical. While Chomsky seems to refer being or not being grammatical, and some questions of punctuation and meaning as well, to spoken English, I think that his real criterion is: a sentence is grammatical if, in reading or saying it aloud with a natural expression and thoughtfully but ingenuously, it is deemed grammatical by a person who speaks it, or perhaps by a person who hears it. Some problems which might plague others may not bother Chomsky because he speaks remarkably well-connected and grammatical English.

Whether or not the rules of grammar can be embodied in a finite-state machine, Chomsky offers persuasive evidence that it is wrong and cumbersome to try to generate a sentence by basing the choice of the next word entirely and solely on words already written down. Rather, Chomsky considers the course of sentence generation to be something of this sort:

We start with one or another of several general forms the sentence might take; for example, a noun phrase followed by a verb phrase. Chomsky calls such a particular form of sentence a *kernel sentence.* We then invoke rules for expanding each of the parts of the kernel sentence. In the case of a noun phrase we may first describe it as an article plus a noun and finally as "the man." In the case of a verb phrase we may describe it as a verb plus an object, the object as an article plus a noun, and, in choosing particular words, as "hit the ball." Proceeding in this way from the kernel sentence, noun phrase plus verb phrase, we arrive at the sentence, "The man hit the ball." At any stage we could have made other choices. By making other choices at the final stages we might have arrived at "A girl caught a cat."

Here we see that the element of choice is not exercised sequentially along the sentence from beginning to end. Rather, we choose an over-all skeletal plan or scheme for the whole final sentence at the start. That scheme or plan is the kernel sentence. Once the kernel sentence has been chosen, we pass on to parts of the kernel sentence. From each part we proceed to the constituent elements of that part and from the constituent elements to the choice of particular words. At each branch of this treelike structure growing from the kernel sentence, we exercise choice in arriving at the particular final sentence, and, of course, we chose the kernel sentence to start with.

Here I have indicated Chomsky's ideas very incompletely and very sketchily. For instance, in dealing with irregular forms of words Chomsky will first indicate the root word and its particular grammatical form, and then he will apply certain obligatory rules in arriving at the correct English form. Thus, in the branching construction of a sentence, use is made both of optional rules, which allow choice, and of purely mechanical, deterministic obligatory rules, which do not.

To understand this approach further and to judge its merit, one must refer to Chomsky's book,[1] and to the references he gives.

Chomsky must, of course, deal with the problem of ambiguous sentences, such as, "The lady scientist made the robot fast while she ate." The author of this sentence, a learned information theo-

[1] Noam Chomsky, *Syntactic Structures,* Mouton and Co., 's-Gravenhage, 1957.

rist, tells me that, allowing for the vernacular, it has at least four different meanings. It is perhaps too complicated to serve as an example for detailed analysis.

We might think that ambiguity arises only when one or more words can assume different meanings in what is essentially the same grammatical structure. This is the case in "he was mad" (either angry or insane) or "the pilot was high" (in the sky or in his cups). Chomsky, however, gives a simple example of a phrase in which the confusion is clearly grammatical. In "the shooting of the hunters," the noun hunters may be either the subject, as in "the growling of lions" or the object, as in "the growing of flowers."

Chomsky points out that different rules of transformation applied to different kernel sentences can lead to the same sequence of grammatical elements. Thus, "the picture was painted by a real artist" and "the picture was painted by a new technique" seem to correspond grammatically word for word, yet the first sentence could have arisen as a transformation of "a real artist painted the picture" while the second could not have arisen as a transformation of a sentence having this form. When the final words as well as the final grammatical elements are the same, the sentence is ambiguous.

Chomsky also faces the problem that the distinction between the provinces of grammar and meaning is not clear. Shall we say that grammar allows adjectives but not adverbs to modify nouns? This allows "colorless green." Or should grammar forbid the association of some adjectives with some nouns, of some nouns with some verbs, and so on? With one choice, certain constructions are grammatical but meaningless; with the other they are ungrammatical.

We see that Chomsky has laid out a plan for a grammar of English which involves at each point in the synthesis of a sentence certain steps which are either obligatory or optional. The processes allowed in this grammar cannot be carried out by a finite-state machine, but they can be carried out by a more general machine called a *Turing machine,* which is a finite-state machine plus an infinitely long tape on which symbols can be written and from which symbols can be read or erased. The relation of Chomsky's grammar to such machines is a proper study for those interested in automata.

We should note, however, that if we arbitrarily impose some bound on the length of a sentence, even if we limit the length to 1,000 or 1 million words, then Chomsky's grammar *does* correspond to a finite-state machine. The imposition of such a limit on sentence length seems very reasonable in a practical way.

Once a general specification or model of a grammar of the sort Chomsky proposes is set up, we may ask under what circumstances and how can an entropy be derived which will measure the choice or uncertainty of a message source that produces text according to the rules of the grammar? This is a question for the mathematically skilled information theorist.

Much more important is the production of a plausible and workable grammar. This might be a *phrase-structure* grammar, as Chomsky proposes, or it might take some other form. Such a grammar might be incomplete in that it failed to produce or analyze some constructions to be found in grammatical English. It seems more important that its operation should correspond to what we know of the production of English by human beings. Further, it should be simple enough to allow the generation and analysis of text by means of an electronic computer. I believe that computers must be used in attacking problems of the structure and statistics of English text.

While a great many people are convinced that Chomsky's phrase-structure approach is a very important aspect of grammar, some feel that his picture of the generation of sentences should be modified or narrowed if it is to be used to describe the actual generation of sentences by human beings. Subjectively, in speaking or listening to a speaker one has a strong impression that sentences are generated largely from beginning to end. One also gets the impression that the person generating a sentence doesn't have a very elaborate pattern in his head at any one time but that he elaborates the pattern as he goes along.

I suspect that studies of the form of grammars and of the statistics of their use as revealed by language will in the not distant future tell us many new things about the nature of language and about the nature of men as well. But, to say something more particular than this, I would have to outreach present knowledge—mine and others.

A grammar must specify not only rules for putting different types of words together to make grammatical structures; it must divide the actual words of English into classes on the basis of the places in which they can appear in grammatical structures. Linguists make such a division purely on the basis of grammatical function without invoking any idea of meaning. Thus, all we can expect of a grammar is the generation of grammatical sentences, and this includes the example given earlier: "The chartreuse semiquaver skinned the feelings of the manifold." Certainly the division of words into grammatical categories such as nouns, adjectives, and verbs is not our sole guide concerning the use of words in producing English text.

What does influence the choice among words when the words used in constructing grammatical sentences are chosen, not at random by a machine, but rather by a live human being who, through long training, speaks or writes English according to the rules of the grammar? This question is not to be answered by a vague appeal to the word *meaning.* Our criteria in producing English sentences can be very complicated indeed. Philosophers and psychologists have speculated about and studied the use of words and language for generations, and it is as hard to say anything entirely new about this as it is to say anything entirely true. In particular, what Bishop Berkeley wrote in the eighteenth century concerning the use of language is so sensible that one can scarcely make a reasonable comment without owing him credit.

Let us suppose that a poet of the scanning, rhyming school sets out to write a grammatical poem. Much of his choice will be exercised in selecting words which fit into the chosen rhythmic pattern, which rhyme, and which have alliteration and certain consistent or agreeable sound values. This is particularly notable in Poe's "The Bells," "Ulalume," and "The Raven."

Further, the poet will wish to bring together words which through their sound as well as their sense arouse related emotions or impressions in the reader or hearer. The different sections of Poe's "The Bells" illustrate this admirably. There is a marked contrast between:

> How they tinkle, tinkle, tinkle,
> In the icy air of night!
> While the stars that oversprinkle

All the heavens, seem to twinkle
In a crystalline delight; . . .

and

Through the balmy air of night
How they ring out their delight!
From the molten-golden notes,
And all in tune,
What a liquid ditty floats . . .

Sometimes, the picture may be harmonious, congruous, and moving without even the trivial literal meaning of this verse of Poe's, as in Blake's two lines:

Tyger, Tyger, burning bright
In the forests of the night . . .

In instances other than poetry, words may be chosen for euphony, but they are perhaps more often chosen for their associations with and ability to excite passions such as those listed by Berkeley: fear, love, hatred, admiration, disdain. Particular words or expressions move each of us to such feelings. In a given culture, certain words and phrases will have a strong and common effect on the majority of hearers, just as the sights, sounds or events with which they are associated do. The words of a hymn or psalm can induce a strong religious emotion; political or racial epithets, a sense of alarm or contempt, and the words and phrases of dirty jokes, sexual excitement.

One emotion which Berkeley does not mention is a sense of understanding. By mouthing commonplace and familiar patterns of words in connection with ill-understood matters, we can associate some of our emotions of familiarity and insight with our perplexity about history, life, the nature of knowledge, consciousness, death, and Providence. Perhaps such philosophy as makes use of common words should be considered in terms of assertion of a reassurance concerning the importance of man's feelings rather than in terms of meaning.

One could spend days on end examining examples of motivation in the choice of words, but we do continually get back to the matter of meaning. Whatever meaning may be, all else seems lost without it. A Chinese poem, hymn, deprecation, or joke will have little effect on me unless I understand Chinese in whatever sense those who know a language understand it.

Though Colin Cherry, a well-known information theorist, appears to object, I think that it is fair to regard meaningful language as a sort of code of communication. It certainly isn't a simple code in which one mechanically substitutes a word for a deed. It's more like those elaborate codes of early cryptography, in which many alternative code words were listed for each common letter or word (in order to suppress frequencies). But in language, the listings may overlap. And one person's code book may have different entries from another's, which is sure to cause confusion.

If we regard language as an imperfect code of communication, we must ultimately refer meaning back to the intent of the user. It is for this reason that I ask, "What do you mean?" even when I have heard your words. Scholars seek the intent of authors long dead, and the Supreme Court seeks to establish the intent of Congress in applying the letter of the law.

Further, if I become convinced that a man is lying, I interpret his words as meaning that he intends to flatter or deceive me. If I find that a sentence has been produced by a computer, I interpret it to mean that the computer is functioning very cleverly.

I don't think that such matters are quibbles; it seems that we are driven to such considerations in connection with meaning if we do regard language as an imperfect code of communication, and as one which is sometimes exploited in devious ways. We are certainly far from any adequate treatment of such problems.

Grammatical sentences do, however, have what might be called a formal meaning, regardless of intent. If we had a satisfactory grammar, a machine should be able to establish the relations between the words of a sentence, indicating subject, verb, object, and what modifying phrases or clauses apply to what other words. The next problem beyond this in seeking such formal meaning in sentences is the problem of associating words with objects, qualities, actions, or relations in the world about us, including the world of man's society and of his organized knowledge.

In the simple communications of everyday life, we don't have much trouble in associating the words that are used with the proper objects, qualities, actions, and relations. No one has trouble with "close the east window" or "Henry is dead," when he hears such a simple sentence in simple, unambiguous surroundings. In a familiar American room, anyone can point out the window; we have closed windows repeatedly, and we know what direction east is. Also, we know Henry (if we don't get Henry Smith mixed up with Henry Jones), and we have seen dead people. If the sentence is misheard or misunderstood, a second try is almost sure to succeed.

Think, however, how puzzling the sentence about the window would be, even in translation, to a shelterless savage. And we can get pretty puzzled ourselves concerning such a question as, is a virus living or dead?

It appears that much of the confusion and puzzlement about the associations of words with things of the world arose through an effort by philosophers from Plato to Locke to give meaning to such ideas as window, cat, or dead by associating them with general ideas or ideal examples. Thus, we are presumed to identify a window by its resemblance to a general idea of a window, to an ideal window, in fact, and a cat by its resemblance to an ideal cat which embodies all the attributes of cattiness. As Berkeley points out, the abstract idea of a (or the ideal) triangle must at once be "neither oblique, rectangle, equilateral, equicrural nor scaleron, but all and none of these at once."

Actually, when a doctor pronounces a man dead he does so on the basis of certain observed *signs* which he would be at a loss to identify in a virus. Further, when a doctor makes a diagnosis, he does not start out by making an over-all comparison of the patient's condition with an ideal picture of a disease. He first looks for such signs as appearance, temperature, pulse, lesions of the skin, inflammation of the throat, and so on, and he also notes such *symptoms* as the patient can describe to him. Particular combinations of signs and symptoms indicate certain diseases, and in differential diagnoses further tests may be used to distinguish among diseases producing similar signs and symptoms.

In a similar manner, a botanist identifies a plant, familiar or unfamiliar, by the presence or absence of certain qualities of size, color, leaf shape and disposition, and so on. Some of these qualities, such as the distinction between the leaves of monocotyledonous and dicotyledonous plants, can be decisive; others, such as size, can be merely indicative. In the end, one is either sure he is right or perhaps willing to believe that he is right; or the plant may be a new species.

Thus, in the workaday worlds of medicine and botany, the ideal disease or plant is conspicuous by its absence as any actual useful criterion. Instead, we have lists of qualities, some decisive and some merely indicative.

The value of this observation has been confirmed strongly in recent work toward enabling machines to carry out tasks of recognition or classification. Early workers, perhaps misled by early philosophers, conceived the idea of matching a letter to an ideal pattern of a letter or the spectrogram of a sound to an ideal spectrogram of the sound. The results were terrible. Audrey, a pattern-matching machine with the bulk of a hippo and brains beneath contempt, could recognize digits spoken by one voice or a selected group of voices, but Audrey was sadly fallible. We should, I think, conclude that human recognition works this way in very simple cases only, if at all.

Later and more sophisticated workers in the field of recognition look for significant features. Thus, as a very simple example, rather than having an ideal pattern of a capital Q, one might describe Q as a closed curve without corners or reversals of curvature and with something attached between four and six o'clock.

In 1959, L. D. Harmon built at the Bell Laboratories a simple device weighing a few pounds which almost infallibly recognizes the digits from one to zero written out as words in longhand. Does this gadget match the handwriting against patterns? You bet it doesn't! Instead, it asks such questions as, how many times did the stylus go above or below certain lines? Were I's dotted or T's crossed?

Certainly, no one doubts that words refer to classes of objects, actions, and so on. We are surrounded by and involved with a large number of classes and subclasses of objects and actions which we can usefully associate with words. These include such objects as plants (peas, sunflowers . . .), animals (cats, dogs . . .), machines (autos, radios . . .), buildings (houses, towers . . .), clothing (skirts, socks . . .), and so on. They include such very complicated sequences of actions as dressing and undressing (the absent-minded, including myself, repeatedly demonstrate that they can do this unconsciously); tying one's shoes (an act which children have considerable difficulty in learning), eating, driving a car, reading, writing, adding figures, playing golf or tennis (activities involving a host of distinct subsidiary skills), listening to music, making love, and so on and on and on.

It seems to me that what delimits a particular class of objects, qualities, actions, or relations is not some sort of ideal example. Rather, it is a list of qualities. Further, the list of qualities cannot be expected to enable us to divide experience up into a set of logical, sharply delimited, and all-embracing categories. The language of science may approach this in dealing with a narrow range of experience, but the language of everyday life makes arbitrary, overlapping, and less than all-inclusive divisions of experience. Yet, I believe that it is by means of such lists of qualities that we identify doors, windows, cats, dogs, men, monkeys, and other objects of daily life. I feel also that this is the way in which we identify common actions such as running, skipping, jumping, and tying, and such symbols as words, written and spoken, as well.

I think that it is only through such an approach that we can hope to make a machine classify objects and experience in terms of language, or recognize and interpret language in terms of other language or of action. Further, I believe that when a word cannot offer a table of qualities or signs whose elements can be traced back to common and familiar experiences, we have a right to be wary of the word.

If we are to understand language in such a way that we can hope some day to make a machine which will use language successfully, we must have a grammar and we must have a way of relating words to the world about us, but this is of course not enough. If we are to regard sentences as meaningful, they must in some way correspond to life as we live it.

Our lives do not present fresh objects and fresh actions each day. They are made up of familiar objects and familiar though complicated sequences of actions presented in different groupings and orders. Sometimes we learn by adding new objects, or actions, or combinations of objects or sequences of actions to our stock, and so we enrich or change our lives. Sometimes we forget objects and actions.

Our particular actions depend on the objects and events about us. We dodge a car (a complicated sequence of actions). When thirsty, we stop at the fountain and drink (another complicated but recurrent sequence). In a packed crowd we may shoulder someone out of the way as we have done before. But our information about the world does not all come from direct observation, and our influence on others is happily not confined to pushing and shoving. We have a powerful tool for such purposes: language and words.

We use words to learn about relations among objects and activities and to remember them, to instruct others or to receive instruction from them, to influence people in one way or another. For the words to be useful, the hearer must understand them in the same sense that the speaker means them, that is, insofar as he associates them with nearly enough the same objects or skills. It's no use, however, to tell a man to read or to add a column of figures if he has never carried out these actions before, so that he doesn't have these skills. It is no use to tell him to shoot the aardvark and not the gnu if he has never seen either.

Further, for the sequences of words to be useful, they must refer to real or possible sequences of events. It's of no use to advise a man to walk from London to New York in the forenoon immediately after having eaten a seven o'clock dinner.

Thus, in some way the meaningfulness of language depends not only on grammatical order and on a workable way of associating words with collections of objects, qualities, and so on; it also depends on the structure of the world around us. Here we encounter a real and an extremely serious difficulty with the idea that we can in some way translate sentences from one language into another and accurately preserve the "meaning."

One obvious difficulty in trying to do this arises from differences in classification. We can refer to either the foot or the lower leg; the Russians have one word for the foot plus the lower leg. Hungarians have twenty fingers (or toes), for the word is the same for either appendage. To most of us today, a dog is a dog, male or female, but men of an earlier era distinguished sharply between a dog and a bitch. Eskimos make, it is said, many distinctions among snow which in our language would call for descriptions, and for us even these descriptions would have little real content of importance or feeling, because in our lives the distinctions have not been important. Thus, the parts of the world which are common and meaningful to those speaking different languages are often divided into somewhat different classes. It may be impossible to write down in different languages words or simple sentences that specify exactly the same range of experience.

There is a graver problem than this, however. The range of experience to which various words refer is not common among all cultures. What is one to do when faced with the problem of translating a novel containing the phrase, "tying one's shoelace," which as we have noted describes a complicated action, into the language of a shoeless people? An elaborate description wouldn't call up the right thing at all. Perhaps some cultural equivalent (?) could be found. And how should one deal with the fact that "he built a house" means personal tree cutting and adzing in a pioneer novel, while it refers to the employment of an architect and a contractor in a contemporary story?

It is possible to make some sort of translation between closely related languages on a word-for-word or at least phrase-for-phrase basis, though this is said to have led from "out of sight, out of mind" to "blind idiot." When the languages and cultures differ in major respects, the translator has to think what the words mean in terms of objects, actions, or emotions and then express this meaning in the other language. It may be, of course, that the culture with which the language is associated has no close equivalents to the objects or actions described in the passage to be translated. Then the translator is really stuck.

How, oh how is the man who sets out to build a translating machine to cope with a problem such as this? He certainly cannot do so without in some way enabling the machine to deal effectively with what we refer to as understanding. In fact, we see understanding at work even in situations which do not involve translation from one language into another. A screen writer who can quite accurately transfer the essentials of a scene involving a dying uncle in Omsk to one involving a dying father in Dubuque will repeatedly make complete nonsense in trying to rephrase a simple technical statement. This is clearly because he understands grief but not science.

Having grappled painfully with the word *meaning,* we are now faced with the word *understanding.* This seems to have two sides. If we understand algebra or calculus, we can use their manipulations to solve problems we haven't encountered before or to supply proofs of theorems we haven't seen proved. In this sense, understanding is manifested by a power to do, to create, not merely to repeat. To some degree, an electronic computer which proves theorems in mathematical logic which it has not encountered before (as computers can be programmed to do) could perhaps be said to understand the subject. But there is an emotional side to understanding, too. When we can prove a theorem in several ways and fit it together with other theorems or facts in various manners, when we can view a field from many aspects and see how it all fits together, we say that we understand the subject deeply. We attain a warm and confident feeling about our ability to cope with it. Of course, at one time or another most of us have felt the warmth without manifesting the ability. And how disillusioned we were at the critical test!

In discussing language from the point of view of information theory, we have drifted along a tide of words, through the imperfectly charted channels of grammar and on into the obscurities of meaning and understanding. This shows us how far ignorance can take one. It would be absurd to assert that information theory, or anything else, has enabled us to solve the problems of linguistics, of meaning, of understanding, of philosophy, of life. At best, we can perhaps say that we are pushing a little beyond the mechanical constraints of language and getting at the amount of choice that language affords. This idea suggests views concerning the use and function of language, but it does not establish them. The reader may share my freely offered ignorance concerning these matters, or he may prefer his own sort of ignorance.

CHAPTER VII *Efficient Encoding*

WE WILL NEVER AGAIN understand nature as well as Greek philosophers did. A general explanation of common phenomena in terms of a few all-embracing principles no longer satisfies us. We know too much. We must explain many things of which the Greeks were unaware. And, we require that our theories harmonize in detail with the very wide range of phenomena which they seek to explain. We insist that they provide us with useful guidance rather than with rationalizations. The glory of Newtonian mechanics is that it has enabled men to predict the positions of planets and satellites and to understand many other natural phenomena as well; it is surely not that Newtonian mechanics once inspired and supported a simple mechanistic view of the universe at large, including life.

Present-day physicists are gratified by the conviction that all (non-nuclear) physical, chemical, and biological properties of matter can in principle be completely and precisely explained in all their detail by known quantum laws, assuming only the existence of electrons and of atomic nuclei of various masses and charges. It is somewhat embarrassing, however, that the only physical system all of whose properties actually have been calculated exactly is the isolated hydrogen atom.

Physicists are able to predict and explain some other physical phenomena quite accurately and many more semiquantitatively. However, a basic and accurate theoretical treatment, founded on electrons, nuclei, and quantum laws only, without recourse to other experimental data, is lacking for most common thermal, mechanical, electrical, magnetic, and chemical phenomena. Tracing complicated biological phenomena directly back to quantum first principles seems so difficult as to be scarcely relevant to the real problems of biology. It is almost as if we knew the axioms of an important field of mathematics but could prove only a few simple theorems.

Thus, we are surrounded in our world by a host of intriguing problems and phenomena which we cannot hope to relate through one universal theory, however true that theory may be in principle. Until recently the problems of science which we commonly associate with the field of physics have seemed to many to be the most interesting of all the aspects of nature which still puzzle us. Today, it is hard to find problems more exciting than those of biochemistry and physiology.

I believe, however, that many of the problems raised by recent advances in our technology are as challenging as any that face us. What could be more exciting than to explore the potentialities of electronic computers in proving theorems or in simulating other behavior we have always thought of as "human"? The problems raised by electrical communication are just as challenging. Accurate measurements made by electrical means have revolutionized physical acoustics. Studies carried out in connection with telephone transmission have inaugurated a new era in the study of speech and hearing, in which previously accepted ideas of physiology, phonetics, and liguistics have proved to be inadequate. And, it is this chaotic and intriguing field of much new ignorance and of a little new knowledge to which communication theory most directly applies.

If communication theory, like Newton's laws of motion, is to be taken seriously, it must give us useful guidance in connection with problems of communication. It must demonstrate that it has a real and enduring substance of understanding and power. As the name implies, this substance should be sought in the efficient and accurate transmission of information. The substance indeed exists. As we have seen, it existed in an incompletely understood form even before Shannon's work unified it and made it intelligible.

To deal with the matter of accurate transmission of information we need new basic understanding, and this matter will be tackled in the next chapter. The foregoing chapters have, however, put us in a position to discuss some challenging aspects of the efficient transmission of information.

We have seen that in the entropy of an information source measured in bits per symbol or per second we have a measure of the number of binary digits, of off-or-on pulses, per symbol or per second which are necessary to transmit a message. Knowing this number of binary digits required for encoding and transmission, we naturally want a means of actually encoding messages with, at the most, not many more binary digits than this minimum number.

Novices in mathematics, science, or engineering are forever demanding infallible, universal, mechanical methods for solving problems. Such methods are valuable in proving that problems can be solved, but in the case of difficult problems they are seldom practical, and they may sometimes be completely unfeasible. As an example, we may note that an explicit solution of the general cubic equation exists, but no one ever uses it in a practical problem. Instead, some approximate method suited to the type or class of cubics actually to be solved is resorted to.

The person who isn't a novice thinks hard about a specific problem in order to see if there isn't some better approach than a machine-like application of what he has been taught. Let us see how this applies in the case of information theory. We will first consider the case of a discrete source which produces a string of symbols or characters.

In Chapter V, we saw that the entropy of a source can be computed by examining the relative probabilities of occurrence of various long blocks of characters. As the length of the block is increased, the approximation to the entropy gets closer and closer. In a particular case, perhaps blocks 5, or 10, or 100 characters in length might be required to give a very good approximation to the entropy.

We also saw that by dividing the message into successive blocks of characters, to each of which a probability of occurrence can be attached, and by encoding these blocks into binary digits by means of the Huffman code, the number of digits used per character approaches the entropy as the blocks of characters are made longer and longer.

Here indeed is our foolproof mechanical scheme. Why don't we simply use it in all cases?

To see one reason, let us examine a very simple case. Suppose that an information source produces a binary digit, a 1 or a 0, randomly and with equal probability and then follows it with the same digit twice again before producing independently another digit. The message produced by such a source might be:

0 0 0 1 1 1 0 0 0 1 1 1 1 1 1 0 0 0 0 0 0 1 1 1

Would anyone be foolish enough to divide such a message successively into blocks of 1, 2, 3, 4, 5, etc., characters, compute the probabilities of the blocks, encode them with a Huffman code, and note the improvement in the number of binary digits required for transmission? I don't know; it sometimes seems to me that there are no limits to human folly.

Clearly, a much simpler procedure is not only adequate but absolutely perfect. Because of the repetition, the entropy is clearly the same as for a succession of a third as many binary digits chosen randomly and independently with equal probability of 1 or 0. That is, it is ⅓ binary digit per character of the repetitious message. And, we can transmit the message perfectly efficiently simply by sending every third character and telling the recipient to write down each received character three times.

This example is simple but important. It illustrates the fact that we should look for natural structure in a message source, for salient features of which we can take advantage.

The discussion of English text in Chapter IV illustrates this. We might, for instance, transmit text merely as a picture by television or facsimile. This would take many binary digits per character. We would be providing a transmission system capable of sending not only English text, but Cyrillic, Greek, Sanskrit, Chinese, and other text, and pictures of landscapes, storms, earthquakes, and Marilyn Monroe as well. We would not be taking advantage of the elementary and all-important fact that English text is made up of letters.

If we encode English text letter by letter, taking no account of the different probabilities of various letters (and excluding the space), we need 4.7 binary digits per letter. If we take into account the relative probabilities of letters, as Morse did, we need 4.14 binary digits per letter.

If we proceeded mechanically to encode English text more efficiently, we might go on to encoding pairs of letters, sequences of three letters, and so on. This, however, would provide for encoding many sequences of letters which aren't English words. It seems much more sensible to go on to the next larger unit of English text, the word. We have seen in Chapter IV that we would expect to use only about 9 binary digits per word or 1.7 binary digits per character in so encoding English text.

If we want to proceed further, the next logical step would be to consider the structure of phrases or sentences; that is, to take advantage of the rules of grammar. The trouble is that we don't know the rules of grammar completely enough to help us, and if we did, a communication system which made use of these rules would probably be impractically complicated. Indeed, in practical cases it still seems best to encode the letters of English text independently, using at least 5 binary digits per character.

It is, however, important to get some idea of what *could* be accomplished in transmitting English text. To this end, Shannon considered the following communication situation. Suppose we ask a man, using all his knowledge of English, to guess what the next character in some English text is. If he is right we tell him so, and he writes the character down. If he is wrong, we may either tell him what the character actually is or let him make further guesses until he guesses the right character.

Now, suppose that we regard this process as taking place at the transmitter, and say that we have an absolutely identical twin to guess for us at the receiver, a twin who makes just the same mistakes that the man at the transmitter does. Then, to transmit the text, we let the man at the receiver guess. When the man at the transmitter guesses right, so will the man at the receiver. Thus, we need send information to the man at the receiver only when the man at the transmitter guesses wrong and then only enough information to enable the men at the transmitter and the receiver to write down the right character.

Shannon has drawn a diagram of such a communication system, which is shown in Figure VII-1. A predictor acts on the original text. The prediction of the next letter is compared with the actual letter. If an error is noted, some information is transmitted. At the receiver, a prediction of the next character is made from the already reconstructed text. A comparison involving the received signal is carried out. If no error has been made, the predicted character is used; if an error has been made, the "reduced text" information coming in will make it possible to correct the error.

Of course, we don't have such identical twins or any other highly effective identical predictors. Nonetheless, a much simpler but purely mechanical system based on this diagram has been used in transmitting pictures. Shannon's purpose was different, however. By using just one person, and not twins, he was able to find what transmission rate would be required in such a system merely by examining the errors made by the one man in the transmitter situation. The results are summed up in Figure V-4 of Chapter V. A better prediction is made on the basis of the 100 preceding letters than on the basis of the preceding 10 or 15. To correct the errors in prediction, something between 0.6 and 1.3 binary digits per character is required. This tells us that, insofar as this result is correct, the entropy of English text must lie between .6 and 1.3 bits per letter.

A discrete source of information provides a good example for discussion but not an example of much practical importance in communication. The reason is that, by modern standards of electrical communication, it takes very few binary digits or off-or-on pulses to send English text. We have to hurry to speak a few hundred words a minute, yet it is easy to send over a thousand words of text over a telephone connection in a minute or to send 10 million words a minute over a TV channel, and, in principle if not in practice, we could transmit some 50,000 words a minute over

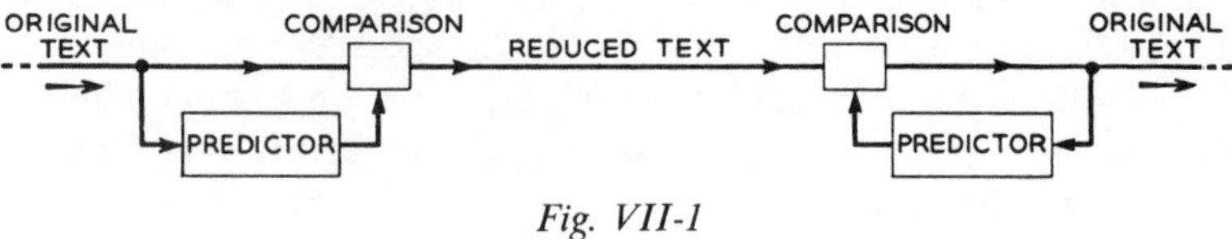

Fig. VII-1

a telephone channel and some 50 million words a minute over a TV channel. As a matter of fact, in practical cases we have even retreated from Morse's ingenious code which sends an E faster than a Z. A teletype system uses the same length of signal for any letter.

Efficient encoding is thus potentially more important for voice transmission than for transmission of text, for voice takes more binary digits per word than does text. Further, efficient encoding is potentially more important for TV than for voice.

Now, a voice or a TV signal is inherently continuous as opposed to English text, numbers, or binary digits, which are discrete. Disregarding capitalization and punctuation, an English character may be any one of the letters or the space. At a given moment, the sound wave or the human voice may have any pressure at all lying within some range of pressures. We have noted in Chapter IV that if the frequencies of such a continuous signal are limited to some bandwidth B, the signal can be accurately represented by $2B$ samples or measurements of amplitude per second.

We remember, however, that the entropy per character depends on how many values the character can assume. Since a continuous signal can assume an infinite number of different values at a sample point, we are led to assume that a continuous signal must have an entropy of an infinite number of bits per sample.

This would be true if we required an absolutely accurate reproduction of the continuous signal. However, signals are transmitted to be heard or seen. Only a certain degree of fidelity of reproduction is required. Thus, in dealing with the samples which specify continuous signals, Shannon introduces a *fidelity criterion.* To reproduce the signal in a way meeting the fidelity criterion requires only a finite number of binary digits per sample or per second, and hence we can say that, within the accuracy imposed by a particular fidelity criterion, the entropy of a continuous source has a particular value in bits per sample or bits per second.

It is extremely important to realize that the fidelity criterion should be associated with long stretches of the signal, not with individual samples. For instance, in transmitting a sound, if we make each sample 10 per cent larger, we will merely make the sound louder, and no damage will be done to its quality. If we make a random error of 10 per cent in each sample, the recovered signal will be very noisy. Similarly, in picture transmission an error in brightness or contrast which changes smoothly and gradually across the picture will pass unnoticed, but an equal but random error differing from point to point will be intolerable.

We have seen that we can send a continuous signal by quantizing each sample, that is, by allowing it to assume only certain preassigned values. It appears that 128 values are sufficient for the transmission of telephone-quality speech or of pictures. We must realize, however, that, in quantizing a speech signal or a picture signal sample by sample, we are proceeding in a very unsophisticated manner, just as we are if we encode text letter by letter rather than word by word.

The name *hyperquantization* has been given to the quantization of continuous signals of more than one sample at a time. This is undoubtedly the true road to efficient encoding of continuous signals. One can easily ruin his chances of efficient encoding completely by quantizing the samples at the start. Yet, to hyperquantize a continuous signal is not easy. Samples are quantized independently in present pulse code modulation systems that carry telephone conversations from telephone office to telephone office and from town to town, and in the digital switching systems that provide much long distance switching. Samples are quantized independently in sending pictures back from Mars, Jupiter and farther planets.

In pulse code modulation, the nearest of one of a number of standard levels or amplitudes is assigned to each sample. As an example, if eight levels were used, they might be equally spaced as in a of Figure VII-2. The level representing the sample is then transmitted by sending the binary number written to the right of it.

Some subtlety of encoding can be used even in such a system. Instead of the equally spaced amplitudes of Figure VII-2*a*, we can use quantization levels which are close together for small signals and farther apart for large signals, as shown in Figure VII-2*b*. The reason for doing this is, of course, that our ears are sensitive to a fractional error in signal amplitude rather than to an error of so many dynes below or above average pressure or so many volts positive or negative, in the signal. By such *companding* (*compressing* the high amplitudes at the transmitter and *expanding* them again

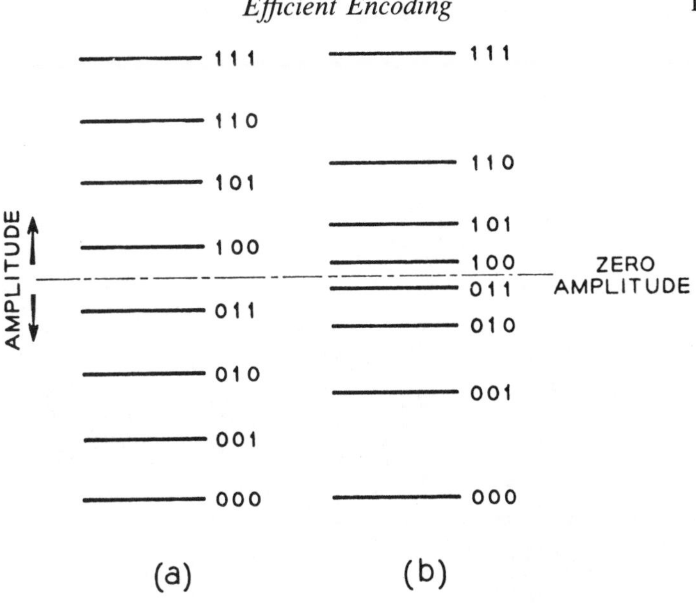

Fig. VII-2

at the receiver), 7 binary digits per sample can give a signal almost as good as 11 binary digits would if the signal levels transmitted were separated by equal differences in amplitude.

To send speech more efficiently than this, we need to examine the characteristics both of speech and of hearing. After all, we require only enough accuracy of transmission to convince the hearer that transmission is good enough.

Efficiency is not everything. A vocoder can transmit only one voice, not two or more at a time. Also, vocoders behave badly when one speaks in the presence of loud noise. Trying to transmit the actual speech waveform more efficiently, or *waveform decoding*, avoids these problems, but 15,000–20,000 binary digits per second are required for acceptable speech.

Figure VII-3 shows the wave forms of several speech sounds, that is, how the pressure of the sound wave or the voltage representing it in a communication system varies with time. We see that many of the wave forms, and especially those for the vowels (*a* through *d*), repeat over and over almost exactly. Couldn't we perhaps transmit just one complete period of variation and use it

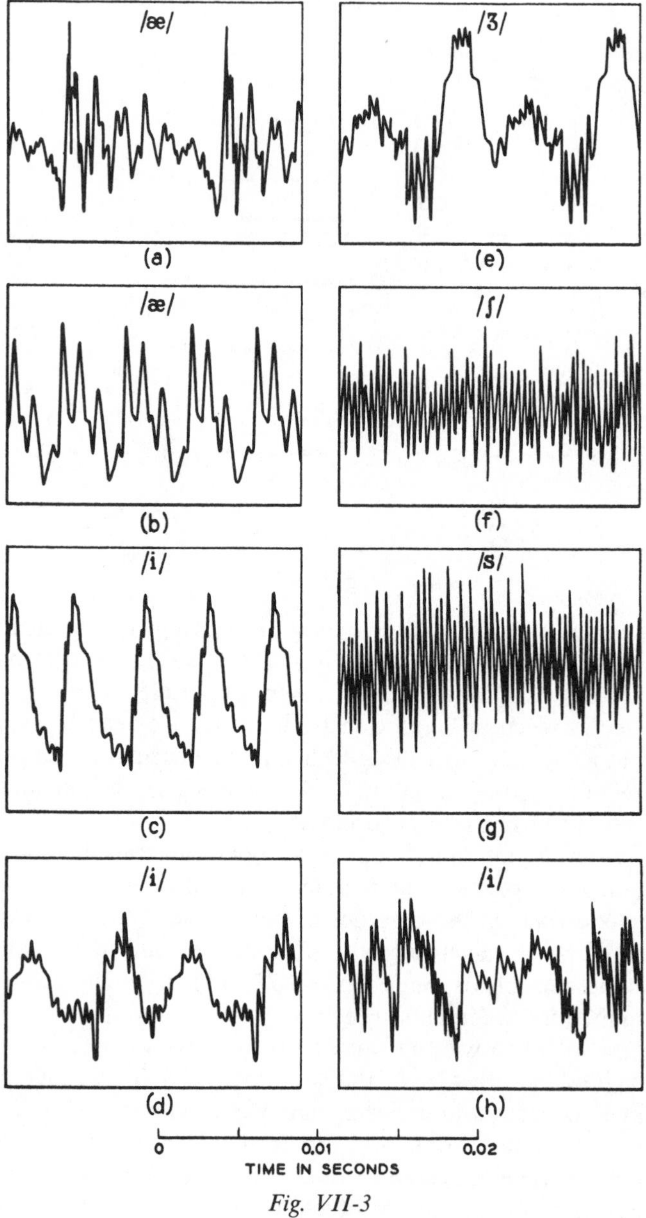

Fig. VII-3

to replace several succeeding periods? This is very difficult, for it is hard for a machine to determine just how long a period is in actual speech. It has been tried. The speech reproduced is intelligible but seriously distorted.

If speech is to be encoded efficiently, a much more fundamental approach is required. We must know how great a variety of speech sounds must be transmitted and how effective our sense of hearing is in distinguishing among speech sounds.

The fluctuations of air pressure which constitute the sounds of speech are very rapid indeed, of the order of thousands per second. Our voluntary control over our vocal tracts is exercised at a much lower rate. At the most, we change the manner of production of sounds a few tens of times a second. Thus, speech may well be (and is) simpler than we might conclude by examining the rapidly fluctuating sound waves of speech.

What control do we exercise over our vocal organs? First of all, we control the production of *voiced* sounds by our control over our vocal cords. These are two lips or folds of muscular tissue attached to a cartilaginous box called the *larynx,* which is prominent in man as the Adam's apple. When we are not giving voice to sound, these are wide open. They can be drawn together more or less tightly, so that when air from the lungs is forced through them they emit a sound something like a Bronx cheer. If they are held very tight, the sound has a high pitch; if they are more relaxed, the sound has a lower pitch.

The pulses of air passing the vocal cords contain many frequencies. The mouth and lips act as a complex resonator which emphasizes certain frequencies more than others. What frequencies are emphasized depends on how much and at what position the tongue is raised or humped in the mouth, on whether the soft palate opens the nasal cavities to the mouth and throat, and on the opening of the jaws and the position of the lips.

Particular sounds of voiced speech, which includes vowels and other *continuants,* such as m and r, are formed by exciting the vocal cords and giving particular characteristic shapes to the mouth.

Stop consonants, or *plosives,* such as p, b, g, t, are formed by stopping off the vocal passage at various points with the tongue or lips, creating an air pressure, and suddenly releasing it. The vocal cords are used in producing some of these sounds (b, for instance) and not in producing others (p, for instance).

Fricatives, such as s and sh, are produced by the passage of air through various constrictions. Sometimes the vocal cords are used as well (in a zh sound, as in azure).

A specification of the movements of the vocal organs would be much more slowly changing than a description of the sound produced. May this not be a clue to efficient encoding of speech?

In the early thirties, long before Shannon's work on information theory, Homer Dudley of the Bell Laboratories invented such a form of speech transmission, which he called the vocoder (from voice coder). The transmitting (analyzer) and receiving (synthesizer) units of a vocoder are illustrated in Figure VII-4.

In the analyzer, an electrical replica of the speech is fed to 16 filters, each of which determines the strength of the speech signal in a particular band of frequencies and transmits a signal to the synthesizer which gives this information. In addition, an analysis is made to determine whether the sound is voiceless (s, f) or voiced (o, u) and, if voiced, what the pitch is.

At the synthesizer, if the sound is voiceless, a hissing noise is produced; if the sound is voiced a sequence of electrical pulses is produced at the proper rate, corresponding to the puffs of air passing the vocal cords of the speaker.

The hiss or pulses are fed to an array of filters, each passing a band of frequencies corresponding to a particular filter in the analyzer. The amount of sound passing through a particular filter in the synthesizer is controlled by the output of the corresponding analyzer filter so as to be the same as that which the analyzer filter indicates to be present in the voice in that frequency range.

This process results in the reproduction of intelligible speech. In effect, the analyzer listens to and analyzes speech, and then instructs the synthesizer, which is an artificial speaking machine, how to say the words all over again with the very pitch and accent of the speaker.

Most vocoders have a strong and unpleasant electrical accent. The study of this has led to new and important ideas concerning what determines and influences speech quality; we cannot afford time to go into this matter here. Even imperfect vocoders can be

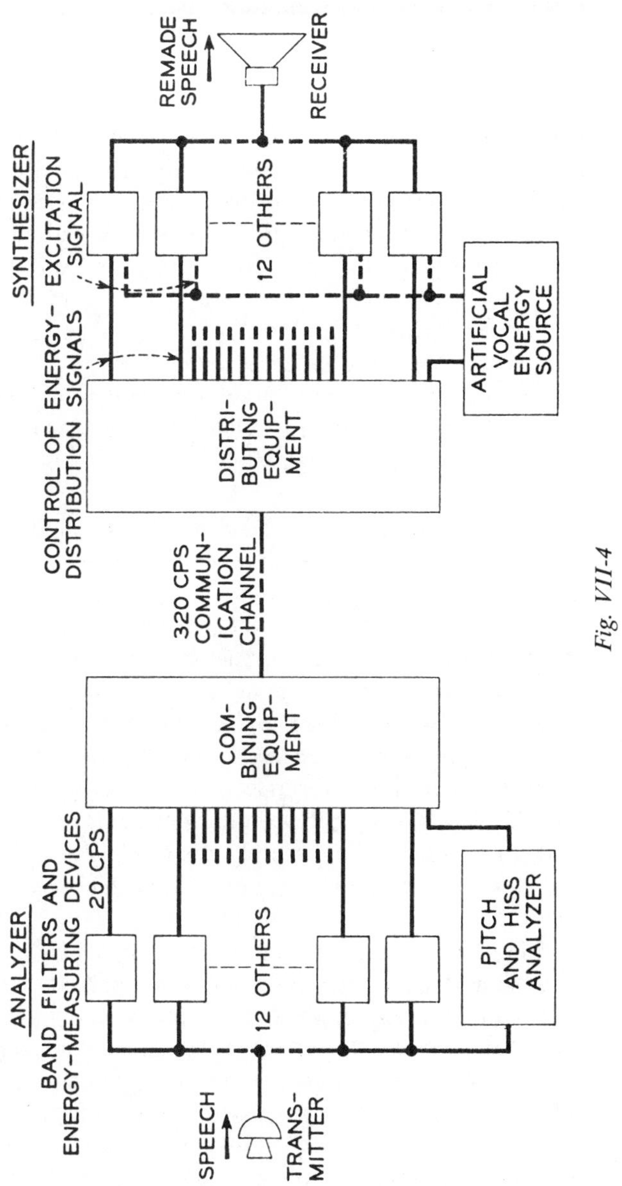
ANALYZER
BAND FILTERS AND
ENERGY-MEASURING DEVICES
20 CPS
12 OTHERS
SPEECH
TRANS-
MITTER
PITCH
AND HISS
ANALYZER
COM-
BINING
EQUIP-
MENT
320 CPS
COMMUN-
ICATION
CHANNEL
DISTRI-
BUTING
EQUIP-
MENT
SYNTHESIZER
CONTROL OF ENERGY-
DISTRIBUTION SIGNALS
EXCITATION
SIGNAL
12 OTHERS
ARTIFICIAL
VOCAL
ENERGY
SOURCE
REMADE
SPEECH
RECEIVER

Fig. VII-4

very useful. For instance, it is sometimes necessary to resort to enciphered speech transmission. If one merely directly reduces speech to binary digits by pulse code modulation, 30,000 to 60,000 binary digits per second must be sent. By using a vocoder, speech can be sent with about 2,400 binary digits per second.

The channel vocoder of Figure VII-4 is only one example of a large class of devices (we may call them all vocoders, if we wish) that analyze speech and transmit signals which drive a speaking machine. In linear predictive encoding the analysis finds slowly varying coefficients that predict the next speech sample as a weighted sum of several past samples. An error signal can be sent as well, which is used to correct the output of the speaking machine. Linear predictive coding gives very good speech if 9,600 binary digits per second are transmitted, intelligible speech at 2,400 binary digits per second, and barely intelligible speech at 600 binary digits per second.

Various other parameters of speech can be derived from the linear predictive coefficients. The channel signals characteristic of the channel vocoder of Figure VII-4 can be derived from the linear predictive coefficients. So can the resonant frequencies of the vocal tract characteristic of various speech sounds. These resonant frequencies are called *formants.* When we transmit these resonant frequencies and use them to reconstruct speech we say we have a *formant tracking vocoder.* It has been proposed to derive parameters describing the shape of the vocal tract and to transmit these. If, only if, we could use the coefficients to recognize speech sounds, or *phonemes*, and merely transmit their labels, we would have a *phoneme vocoder* that would transmit speech with the efficiency of text.

Let us consider the vocoder for a moment before leaving it.

We note that transmission of voice using even the most economical of vocoders takes many more binary digits per word than transmission of English text. Partly, this is because of the technical difficulties of analyzing and encoding speech as opposed to print. Partly, it is because, in the case of speech, we are actually transmitting information about speech quality, pitch, and stress, and accent as well as such information as there is in text. In other words, the entropy of speech is somewhat greater per word than the entropy of text.

That the vocoder does encode speech more efficiently than other methods depends on the fact that the configuration of the vocal tract changes less rapidly than the fluctuations of the sound waves which the vocal tract produces. Its effectiveness also depends on limitations of the human sense of hearing.

From an electrical point of view, the most complicated speech sounds are the hissing fricatives, such as sh (*f* of Figure VII-3) and s (*g* of Figure VII-3). Furthermore, the wave forms of two s's uttered successively may have quite a different sequence of ups and downs. It would take many binary digits per second to transmit each in full detail. But, to the ear, one s sounds just like another if it has in a broad way the same frequency content. Thus, the vocoder doesn't have to reproduce the s sound the speaker uttered; it has merely to reproduce an s sound that has roughly the same frequency content and hence sounds the same.

We see that, in transmitting speech, the royal road to efficient encoding appears to be the detection of certain simple and important patterns and their recreation at the receiving end. Because of the greater channel capacity required, efficient encoding is even more important in TV transmission than in speech transmission. Can we perhaps apply a similar principle in TV?

The TV problem is much more difficult than the speech transmission problem. Partly, this is because the sense of sight is inherently more detailed and discriminating than the sense of hearing. Partly, though, it is because many sorts of pictures from many sources are transmitted by TV, while speech is all produced by the same sort of vocal apparatus.

In the face of these facts, is some vocoder-like way of transmitting pictures possible if we confine ourselves to one sort of picture source, for instance, the human face?

One can conceive of such a thing. Imagine that we had at the receiver a sort of rubbery model of a human face. Or we might have a description of such a model stored in the memory of a huge electronic computer. First, the transmitter would have to look at the face to be transmitted and "make up" the model at the receiver in shape and tint. The transmitter would also have to note the sources of light and reproduce these in intensity and direction at the receiver. Then, as the person before the transmitter talked, the transmitter would have to follow the movements of his eyes, lips and jaws, and other muscular movements and transmit these so that the model at the receiver could do likewise. Such a scheme might be very effective, and it could become an important invention if anyone could specify a useful way of carrying out the operations I have described. Alas, how much easier it is to say what one would like to do (whether it be making such an invention, composing Beethoven's tenth symphony, or painting a masterpiece on an assigned subject) than it is to do it.

In our day of unlimited science and technology, people's unfulfilled aspirations have become so important to them that a special word, popular in the press, has been coined to denote such dreams. That word is *breakthrough.* More rarely, it may also be used to describe something, usually trivial, which has actually been accomplished.

If we turn from such dreams of the future, we find that all actual picture-transmission systems follow a common pattern. The picture or image to be transmitted is *scanned* to discover the brightness at successive points. The scanning is carried out along a sequence of closely spaced lines. In color TV, three images of different colors are scanned simultaneously. Then, at the receiver, a point of light whose intensity varies in accord with the signal from the transmitter paints out the picture in light and shade, following the same line pattern. So far all practical attempts at efficient encoding have started out with the signal generated by such a scanning process.

The outstanding efficient encoding scheme is that used in color TV. The brightness of a color TV picture has very fine detail; the pattern of color has very much less detail. Thus, color TV of almost the same detail as monochrome TV can be sent over the same channel as is used for monochrome. Of course, color TV uses an analog signal; the picture is not reduced to discrete on-or-off pulses.

Increasingly, pulse code modulation will be used to transmit all sorts of signals, including television signals. The picture to be transmitted will be scanned in a conventional way, but its brightness will be encoded as a succession of binary numbers that specify the brightnesses of a succession of discrete picture elements or pixels that lie along each scanning line. This is how pictures were sent back from Mars by the Mariner lander, and from Jupiter and its moons by the Voyager spacecraft.

All recent work aimed at encoding television efficiently is digital. It deals with successions of binary numbers that represent successive pixel brightnesses.

In large parts of a TV picture the brightness changes gradually and smoothly from pixel to pixel. In such areas of the picture, a good prediction can be made of the brightness of the next pixel from the brightness of preceding pixels in the same line, and perhaps in the preceding line. At the receiver we need know only the error in such a prediction, so we need transmit only the small difference between the true brightness and a brightness which we predict at the receiver as well as at the transmitter. Of course, in "busy" portions of the picture, prediction will be poor, and the brightness difference that must be sent will be great.

We can transmit brightness differences most efficiently by using a Huffman code, with short code words for more frequently occurring small brightness differences and long code words for less frequently occurring, large brightness differences. If we do this, the binary digits of the coded differences will be generated at an uneven rate, at a slow rate when smooth portions of the picture are scanned and at a faster rate when busy portions of the picture are scanned. In order to transmit the binary digits at a constant rate, the digits must be fed into a *buffer*, which stores the incoming digits and feeds them out at a constant rate equal to the average rate at which they come in. A similar buffer must be used at the receiving end.

By means of such *intraframe* encoding, the number of binary digits per second needed to transmit a good TV picture can be reduced to ½ to ⅓ of the number of binary digits used in initially encoding the pixel brightnesses.

Much greater gains can be made through *interframe* encoding, in which the pixel brightnesses of the whole previous TV picture are stored and used in predicting the brightness of the next pixel to be sent. This is particularly effective in transmitting pictures of people against a fixed background, for the brightnesses of pixels in the background don't change from frame to frame.

Even more elaborate experimental schemes make use of the fact that when a figure in front of a background moves, it moves as a whole. Thus, the brightnesses of the pixels in the moving figure can be predicted from the brightnesses of pixels which are a constant distance away in the previous frame.

If each pixel of a TV picture is represented by 8 binary digits (a very good picture), the picture can be transmitted by sending around 100 million binary digits per second. By intraframe encoding this can be reduced to perhaps 32 million. With interframe coding this has been reduced to as little as 6 million. A reduction to 1.5 million seems conceivable for such pictures as the head of a person against a fixed background.

The *transform* method is another approach to the efficient transmission of TV pictures. In the transform method, the pattern of pixel brightnesses that make up the TV picture, or some portion of it, is represented as the sum of a chosen set of standardized patterns whose amplitudes are transmitted with chosen accuracies.

Reviewing what has been said, we see that there are three important principles in encoding signals efficiently: (1) Don't encode the signal one sample or one character at a time; encode a considerable stretch of a signal at a time (hyperquantization); (2) take into account the limitations on the source of the signal; (3) take into account any inabilities of the eye or the ear to detect errors in a reconstruction of the signal.

The vocoder illustrates these principles excellently. The fine temporal structure of the speech wave is not examined in detail. Instead, a description specifying the average intensities over certain ranges of frequencies is transmitted, together with a signal which tells whether the speech is voiced or unvoiced and, if it is voiced, what its pitch is. This description of a signal is efficient because the vocal organs don't change position rapidly in producing speech. At the receiver, the vocoder generates a speech signal which doesn't resemble the original speech signal in fine detail but sounds like the original speech signal, because of the natural limitations of our hearing.

The vocoder is a sort of paragon of efficient transmission devices. Next perhaps comes color TV, in which the variations of color over the picture are defined much less sharply than variations of intensity are. This takes advantage of the eyes' inability to see fine detail in color patterns.

Beyond this, the present art of communication has had to make use of means which, because they do not encode long stretches of signal at a time, must, according to communication theory, be rather inefficient.

Still, efficient encoding is potentially important. This is especially so in the case of the transmission of relatively broad-band signals (TV or even voice signals) over very expensive circuits, such as transoceanic telephone cables.

No doubt much ingenuity will be spent in efficient encoding in the future, and many startling results will be attained. But we should perhaps beware of going too far.

Imagine, for instance, that we send English text letter by letter. If we make an error in sending a few letters we can still make some sense out of the text:

Hore I hove reploced a few vowols by o.

We can even replace the vowels by x's and read with some facility:

Hxrx X hxvx rxplxcxd thx vxwxls bx x.

It is more efficient to encode English text word by word. In this case, if an error is made in transmission, we are not tipped off by finding a misspelled word. Instead, one word is replaced by another. This might have embarrassing results. Suppose it changed "The President is a good Republican" to "The President is a good Communist" (or donkey, or poltroon, or many other nouns).

We might still detect an error by the fact that the word was inappropriate. But suppose we used a more refined encoding scheme that could reproduce grammatical utterances only. Then we would have little chance of detecting an error in transmission.

English text, and most other information sources are *redundant* in that the messages they produce give many clues to the recipient. A few errors caused by replacing one letter by another don't destroy the message because we can infer it from other letters which are transmitted correctly. Indeed, it is only because of this redundancy that anyone can read my handwriting. When a continuous signal is sent a sample at a time, a few errors in sample amplitude result in a few clicks in sound transmission or in a few specks in picture transmission.

Our ideal so far has been to remove this redundancy, so that we transmit the absolutely minimum number of clues by means of which the message can be reconstructed. But we see that if we do this with perfect success, any error in transmission will send, not a distorted message, but a false and misleading message. If we fall a little short of the ideal, an error may produce merely a terrible garble.

We all know that there is some noise in electrical communication —a hiss in the background on radio and a little snow at least in TV. That such noise is an inevitable fact of nature we must accept. Is this going to vitiate in principle our grand plan to encode the messages from a signal source into scarcely more binary digits than the entropy of the source?

This is the subject that we will consider in the next chapter.

CHAPTER VIII *The Noisy Channel*

IT IS HARD TO PUT ONESELF in the place of another, and, especially, it is hard to put oneself in the place of a person of an earlier day. What would a Victorian have thought of present-day dress? Were Newton's laws of motion and of gravitation as astonishing and disturbing to his contemporaries as Einstein's theory of relativity appears to have been to his? And what is disturbing about relativity? Present-day students accept it, not only without a murmur, but with a feeling of inevitability, as if any other idea must be very odd, surprising, and inexplicable.

Partly, this is because our attitudes are bred of our times and surroundings. Partly, in the case of science at least, it is because ideas come into being as a response to new or better-phrased questions. We remember that according to Plato, Socrates drew a geometrical proof from a slave simply by means of an ingenious sequence of questions. Those who have not seriously asked themselves a particular question are not likely to have come upon the proper answer, and, sometimes, when the question is phrased with the answer in mind, the answer appears to be obvious.

Those interested in communication have been aware from the very beginning that communication circuits or channels are imperfect. In telephony and radio, we hear the desired signal against a background of noise, which may be strong or faint and which may vary in quality from the crackling of static to a steady hiss.

In TV, the picture is overlaid faintly or strongly with an ever-changing granular "snow." In teletypewriter transmission, the received character may occasionally differ from that transmitted.

Suppose that one had questioned a communication engineer about this general problem of "noise" in 1945. One might have asked, "What can one *do* about noise?" The engineer might have answered, "You can increase the transmitter power or make the receiver less noisy. And be sure that the receiver is insensitive to disturbances with frequencies other than the signal frequencies."

One might have persisted, "Can't one do anything else?" The engineer might have answered, "Well, by using frequency modulation, which takes a very large band width, one can reduce the effect of noise."

Suppose, however, that one had asked, "In teletypewriter systems, noise may cause some received characters to be wrong; how can one guard against this?" The engineer could and might perhaps have answered, "I know that if I use five off-or-on pulses to represent a decimal digit and assign to the decimal digits only such sequences as all have two ons and three offs, I can often tell when an error has been made in transmission, for when errors are made the received sequence may have other than 2 ons."

One might have pursued the matter further with, "If the teletypewriter circuit does cause errors is there any way that one can get the correct message to the destination?" The engineer might have answered, "I suppose you can if you repeat it enough times, but that's very wasteful. You'd better fix the circuit."

Here we are getting pretty close to questions that just hadn't been asked before Shannon asked them. Nonetheless, let us go on and imagine that one had said, "Suppose that I told you that by properly encoding my message, I can send it over even a noisy channel with a completely negligible fraction of errors, a fraction smaller than any assignable value. Suppose that I told you that, if the sort of noise in the channel is known and if its magnitude is known, I can calculate just how many characters I can send over the channel per second and that, if I send any number fewer than this, I can do so virtually without error, while if I try to send more, I will be bound to make errors."

The engineer might well have answered, "You'd sure have to

show me. I never thought of things in quite that way before, but what you say seems extremely improbable. Why, every time the noise increases, the error rate increases. Of course, repeating a message several times does work better when there aren't too many errors. But, it is always very costly. Maybe there's something in what you say, but I'd be awfully surprised if there was. Still, the way you put it . . ."

Whatever we may imagine concerning an engineer benighted in the days of error, mathematicians and engineers who have survived the transition all feel that Shannon's results concerning the transmission of information over a noisy channel were and still are very surprising. Yet I have known an intelligent layman to see nothing remarkable in Shannon's results. What is one to think of this?

Perhaps the best course is merely to describe and explain the problem of the noisy channel as we *now* understand it, raising and answering questions that, however natural and inevitable they now seem, belong in their trend and content to the post-Shannon era. The reader can be surprised or not as he chooses.

So far we have discussed both simple and complex means for encoding text and numbers for efficient transmission. We have noted further that any electrical signal of limited band width W can be represented by $2W$ amplitudes or samples per second, measured or taken at intervals $1/2W$ seconds apart. We have seen that, by means of pulse code modulation, we can use some number, around 7, of binary digits to represent adequately the amplitude of any sample. Thus, by using pulse code modulation or some more complicated and more efficient scheme, we can transmit speech or picture signals by means of a sequence of binary digits or off-or-on or positive-or-negative pulses of current.

All of this works perfectly if the recipient of the message receives the same signal that the sender transmits. The actual facts are different. Sometimes he receives a 0 when a 1 is transmitted, and sometimes he receives a 1 when a 0 is transmitted. This can happen through the malfunction of electrical relays in a slow-speed telegraph circuit or through the malfunction of vacuum tubes or transistors in a higher speed circuit. It can also happen because of interfering signals or noise, either noise from man-made apparatus, or noise from magnetic storms.

We can easily see in a simple case how errors can occur because of the admixture of noise with a signal. Imagine that we want to send a large number of binary digits, 0 or 1, per second over a wire by means of an electrical signal. We may represent the signal conveying these digits by the succession of samples *s* of Figure VIII-1, each of which will be +1 or −1. Here we have a succession of positive and negative voltages which represent the digits 1 0 1 1 1 0 0 1 0.

Now suppose a random noise voltage, which may be either positive or negative, is added to the signal. We can represent this also by a number of noise samples *n* of Figure VIII-1 taken simultaneously with the signal samples. The signal plus the noise is obtained by adding the signal and the noise samples and is shown as $s + n$ in Figure VIII-1.

If we interpret a positive signal-plus-noise in the received message as a 1 and a negative signal-plus-noise as a 0, then the received

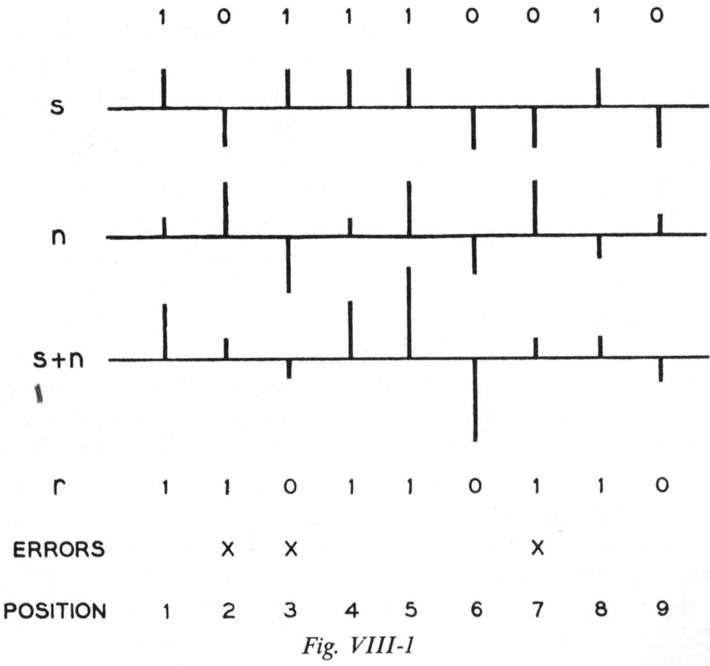

Fig. VIII-1

message will be represented by the digits *r* of Figure VIII-1. Thus, errors in transmission, as indicated, occur in positions 2, 3, and 7.

The effect of such errors in transmission can range from annoying to dangerous. In speech or picture transmission by means of simple coding schemes, they result in clicks, hissing noises, or "snow." If more efficient, block encoding schemes are used (hyperquantization) the effects of errors will be more pronounced. In general, however, we may expect the most dangerous effects of errors in the transmission of text.

In the transmission of English text by conventional means, errors merely put a wrong letter in here and there. The text is so redundant that we catch such errors by eye. However, when type is set remotely by teletypewriter signals, as it is, for instance, in the simultaneous printing of news magazines in several parts of the country, even errors of this sort can be costly.

When numbers are sent errors are much more serious. An error might change $1,000 into $9,000. If the error occurred in a program intended to make an electronic computer carry out a complicated calculation, the error could easily cause the whole calculation to be meaningless.

Further, we have seen that, if we encode English text or any other signal very efficiently, so as largely to remove the redundancy, an error can cause a gross change in the meaning of the received signal.

When errors are very important to us, how indeed may we guard against them? One way would be to send every letter twice or to send every binary digit used in transmitting a letter or a number twice. Thus, in transmitting the binary sequence 1 0 1 0 0 1 1 0 1, we might send and receive as follows:

sent 1 1 0 0 1 1 0 0 0 0 1 1 1 1 0 0 1 1
received 1 1 0 0 1 1 0 0 0 1 1 1 1 1 0 0 1 1
× error

For a given rate of sending binary digits, this will cut our rate of transmitting information in half, for we have to pause and retransmit every digit. However, we can now see from the received signal than an error has occurred at the marked point, because instead of a pair of like digits, 0 0 or 1 1, we have received a pair of unlike digits, 0 1. We don't know whether the correct, transmitted pair was 0 0 or 1 1. We have *detected* the error, but we have not *corrected* it.

If errors aren't too frequent, that is, if the chance of two errors occurring in the transmission of three successive digits is negligible, we can correct as well as detect an error by transmitting each digit three times, as follows:

sent	1 1 1 0 0 0 1 1 1 0 0 0 0 0 0 1 1 1 1 1 1
received	1 1 1 0 0 0 1 0 1 0 0 0 0 0 0 1 1 1 1 1 1
	∧ error

We have now cut our rate of transmission to one-third, because we have to pause and retransmit each digit twice. However, we can now correct the error indicated by the fact that the digits in the indicated group 1 0 1 are not all the same. If we assume that there was only one error in the transmission of this group of digits, then the transmitted group must have been 1 1 1, representing 1, rather than 0 0 0, representing 0.

We see that a very simple scheme of repeating transmitted digits can detect or even correct infrequent errors of transmission. But how costly it is! If we use this means of error correction or detection, even when almost all of the transmitted digits are correct we have to cut our rate of transmission in half by repeating digits in order just to detect errors, and we have to cut our rate of transmission to one-third by transmitting each digit three times in order to get error correction. Moreover, these schemes won't work if errors are frequent enough so that more than one will sometimes occur in the transmission of two or three digits.

Clearly, this simple approach will never lead to a sound understanding of the possibility of error correction. What is required is a deep and powerful mathematical attack. This is just what Shannon provided in discovering and proving his fundamental theorem for the noisy channel. It is the course of his reasoning that we are about to follow.

In formulating an abstract and general model of noise or errors, we will deal with the case of a discrete communication system which transmits some group of characters, such as the digits from 0 to 9 or the letters of the alphabet. For convenience, let us consider a system for transmitting the digits 0 through 9. This is illustrated in Figure VIII-2. At the left we have a number of little circles labeled with the digits; we may regard these little circles as push-buttons. To *he right we have a number of little circles, again labeled with the digits. We may regard these as lights. When we push a digit button at the transmitter to the left, some digit light lights up at the receiver to the right.

If our communication system were noiseless, pushing the 0 button would always light the 0 light, pushing the 1 button would always light the 1 light, and so on. However, in an imperfect or noisy communication system, pushing the 4 button, for instance, may light the 0 light, or the 1 light, or the 2 light, or any other light, as shown by the lines radiating from the 4 button in Figure VIII-2. In a simple, noisy communication system, we can say that when we press a button the light which lights is a matter of chance,

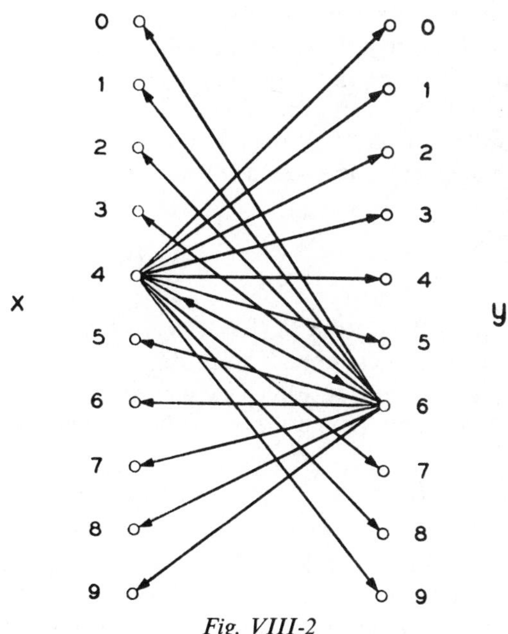

Fig. VIII-2

independent of what has gone before and that, if the 4 button is pressed, there is some probability $p_4(6)$ that the 6 light will light, and so on.

If the sender can't be sure which light will light when he presses a particular button, then the recipient of the message can't be sure which button was pressed when a particular light lights. This is indicated by the arrows from light 6 to various buttons on the left. If, for instance, light 6 lights, there is some probability $p_6(4)$ that button 4 was pressed, and so on. Only for a noiseless system will $p_6(6)$ be unity and $p_6(4)$, $p_6(9)$, etc., be zero.

The diagram of Figure VIII-2 would be too complicated if all possible arrows were put in, and the number of probabilities is too great to list, but I believe that the general idea of the degree and nature of uncertainty of the character received when the sender tries to send a particular character and the uncertainty of the character sent when the recipient receives a particular character, have been illustrated. Let us now consider this noisy communication channel in a rather general way. In doing so we will represent by x all of the characters sent and by y all of the characters received.

The characters x are just the characters generated by the message source from which the message comes. If there are m of these characters and if they occur independently with probabilities $p(x)$, then we know from Chapter V that the entropy $H(x)$ of the message source, the rate at which the message source generates information, must be

$$H(x) = \sum_{x=1}^{m} - p(x) \log p(x) \tag{8.1}$$

We can regard the output of the device, which we designate by y, as another message source. The number of lights need not be equal to the number of buttons, but we will assume that it is, so that there are m lights. The entropy of the output will be

$$H(y) = \sum_{y=1}^{m} - p(y) \log p(y) \tag{8.2}$$

We note that while $H(x)$ depends only on the input to the communication channel, $H(y)$ depends both on the input to the channel and on the errors made in transmission. Thus, the probability of receiving a 4 if nothing but a 4 is ever sent is different from the probability of receiving a 4 if transmitting buttons are pressed at random.

If we imagine that we can see both the transmitter and the receiver, we can observe how often certain combinations of x and y occur; say, how often 4 is sent and 6 is received. Or, knowing the statistics of the message source and the statistics of the noisy channel, we can compute such probabilities. From these we can compute another entropy.

$$H(x, y) = \sum_{x=1}^{m} \sum_{x=1}^{m} - p(x, y) \log p(x, y) \qquad (8.3)$$

This is the uncertainty of the combination of x and y.

Further, we can say, suppose that we know x (that is, we know what key was pressed). What are the probabilities of various lights lighting (as illustrated by the arrows to the right in Figure VIII-2)? This leads to an entropy,

$$H_x(y) = \sum_{x=1}^{m} \sum_{y=1}^{m} - p(x) p_x(y) \log p_x(y) \qquad (8.4)$$

This is a *conditional* entropy of uncertainty. Its form is reminiscent of the entropy of a finite-state machine. As in that case, we multiply the uncertainty for a given condition (state, value of x) by the probability that that condition (state, value of x) will occur and sum over all conditions (states, values of x).

Finally, suppose we know what light lights. We can say what the probabilities are that various buttons were pressed. This leads to another conditional entropy

$$H_y(x) = \sum_{y=1}^{m} \sum_{x=1}^{m} - p(y) p_y(x) \log p_y(x) \qquad (8.5)$$

This is the sum over y of the probability that y is received times the uncertainty that x is sent when y is received.

These conditional entropies depend on the statistics of the message source, because they depend on how often x is transmitted or how often y is received, as well as on the errors made in transmission.

The entropies listed above are best interpreted as uncertainties involving the characters generated by the message source and the characters received by the recipient. Thus:

$H(x)$ is the uncertainty as to x, that is, as to which character will be transmitted.

$H(y)$ is the uncertainty as to which character will be received in the case of a given message source and a given communication channel.

$H(x, y)$ is the uncertainty as to when x will be transmitted and y received.

$H_x(y)$ is the uncertainty of receiving y when x is transmitted. It is the average uncertainty of the sender as to what will be received.

$H_y(x)$ is the uncertainty that x was transmitted when y is received. It is the average uncertainty of the message recipient as to what was actually sent.

There are relations among these quantities:

$$H(x, y) = H(x) + H_x(y) \tag{8.6}$$

That is, the uncertainty of sending x and receiving y is the uncertainty of sending x plus the uncertainty of receiving y when x is sent.

$$H(x, y) = H(y) + H_y(x) \tag{8.7}$$

That is, the uncertainty of receiving y and sending x is the uncertainty of receiving y plus the uncertainty that x was sent when y was received.

We see that when $H_x(y)$ is zero, $H_y(x)$ must be zero, and $H(y)$ is then just $H(x)$. This is the case of the noiseless channel, for which the entropy of the received signal is just the same as the entropy of the transmitted signal. The sender knows just what will be received, and the recipient of the message knows just what was sent.

The uncertainty as to which symbol was transmitted when a given symbol is received, that is, $H_y(x)$ seems a natural measure of the information lost in transmission. Indeed, this proves to be the case, and the quantity $H_y(x)$ has been given a special name; it is called the *equivocation* of the communication channel. If we take $H(x)$ and $H_y(x)$ as entropies in bits per second, the rate R of transmission of information over the channel can be shown to be, in bits per second,

$$R = H(x) - H_y(x) \tag{8.8}$$

That is, the rate of transmission of information is the source rate or entropy less the equivocation. It is the entropy of the message as sent less the uncertainty of the recipient as to what message was sent.

The rate is also given by

$$R = H(y) - H_x(y) \tag{8.9}$$

That is, the rate is the entropy of the received signal y less the uncertainty that y was received when x was sent. It is the entropy of the message as received less the sender's uncertainty as to what will be received.

The rate is also given by

$$R = H(x) + H(y) - H(x, y) \tag{8.10}$$

The rate is the entropy of x plus the entropy of y less the uncertainty of occurrence of the combination x and y. We will note from 8.3 that for a noiseless channel, since $p(x, y)$ is zero except when $x = y$, and $H(x, y) = H(x) = H(y)$. The information rate is just the entropy of the information source, $H(x)$.

Shannon makes expression 8.8 for the rate plausible by means of the sketch shown in Figure VIII-3. Here we assume a system in which an observer compares transmitted and received signals and then sends correction data by means of which the erroneous received signal is corrected. Shannon is able to show that in order to correct the message, the entropy of the correction signal must be equal to the equivocation.

We see that the rate R of relation 8.8 depends both on the channel and on the message source. How can we describe the *capacity* of a noisy or imperfect channel for transmitting information? We can choose the message source so as to make the rate R *as large as possible* for a given channel. This maximum possible rate of transmission for the channel is called the *channel capacity*

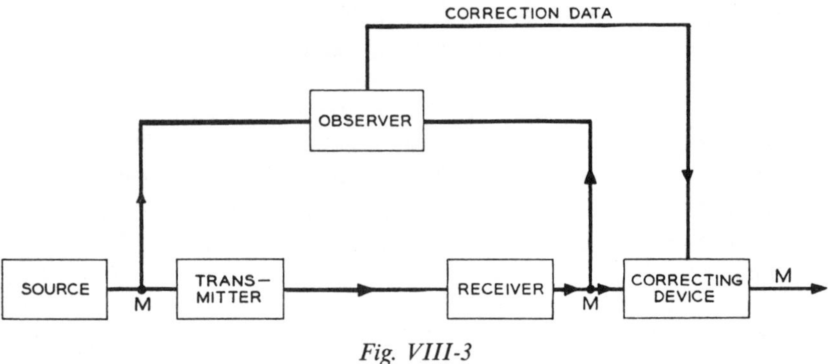

Fig. VIII-3

C. Shannon's fundamental theorem for a noisy channel involves the channel capacity *C.* It says:

> Let a discrete channel have a capacity C and a discrete source the entropy per second H. If H $<$ C there exists a coding system such that the output of the source can be transmitted over the channel with an arbitrarily small frequency of errors (or an arbitrarily small equivocation). If H $>$ C it is possible to encode the source so that the equivocation is less than H $-$ C $+$ ε, where ε is arbitrarily small. There is no method of encoding which gives an equivocation less than H $-$ C.

This is a precise statement of the result which so astonished engineers and mathematicians. As errors in transmission become more probable, that is, as they occur more frequently, the channel capacity as defined by Shannon gradually goes down. For instance, if our system transmits binary digits and if some are in error, the channel capacity *C,* that is, number of bits of information we can send per binary digit transmitted, decreases. But the channel capacity decreases *gradually* as the errors in transmission of digits become more frequent. To achieve transmission with as few errors as we may care to specify, we have to reduce our rate of transmission so that it is equal to or less than the channel capacity.

How are we to achieve this result? We remember that in efficiently encoding an information source, it is necessary to lump many characters together and so to encode the message a long block of characters at a time. In making very efficient use of a noisy channel, it is also necessary to deal with sequences of received characters, each many characters long. Among such blocks, only certain transmitted and received sequences of characters will occur with other than a vanishing probability.

In proving the fundamental theorem for a noisy channel, Shannon finds the average frequency of error for all possible codes (for all associations of particular input blocks of characters with particular output blocks of characters), when the codes are chosen at random, and he then shows that when the channel capacity is greater than the entropy of the source, the error rate averaged over all of these encoding schemes goes to zero as the block length is made very long. If we get this good a result by averaging over all codes chosen at random, then there must be some one of the codes which gives this good a result. One information theorist has characterized this mode of proof as weird. It is certainly not the sort of attack that would occur to an uninspired mathematician. The problem isn't one which would have occurred to an uninspired mathematician, either.

The foregoing work is entirely general, and hence it applies to all problems. I think it is illuminating, however, to return to the example of the binary channel with errors, which we discussed early in this chapter and which is illustrated in Figure VIII-1, and see what Shannon's theorem has to say about this simple and common case.

Suppose that the probability that over this noisy channel a 0 will be received as a 0 is equal to the probability p that a 1 will be received as a 1. Then the probability that a 1 will be received as a 0 or a 0 as a 1 must be $(1 - p)$. Suppose further that these probabilities do not depend on past history and do not change with time. Then, the proper abstract representation of this situation is a symmetric binary channel (in the manner of Figure VIII-2) as shown in Figure VIII-4.

Because of the symmetry of this channel, the maximum information rate, that is, the channel capacity, will be attained for a message source such that the probability of sending a 1 is equal to the probability of sending a zero. Thus, in the case of x (and, because the channel is symmetrical, in the case of y also)

$$p(1) = p(0) = 1/2$$

We already know that under these circumstances

$$\begin{aligned} H(x) &= H(y) \\ &= -\ (\tfrac{1}{2} \log \tfrac{1}{2} + \tfrac{1}{2} \log \tfrac{1}{2}) \\ &= 1 \text{ bit per symbol} \end{aligned}$$

What about the conditional probabilities? What about the equivocation, for instance, as given by 8.5? Four terms will contribute to this conditional entropy. The sources and contributions are:

The probability that 1 is received is ½. When 1 is received, the probability that 1 was sent is p and the probability that 0 was sent is $(1 - p)$. The contribution to the equivocation from these events is:

$$\tfrac{1}{2}(-p \log p - (1 - p) \log (1 - p))$$

There is a probability of ½ that 0 is received. When 0 is received, the probability that 0 was sent is p and the probability that 1 was sent is $(1 - p)$. The contribution to the equivocation from these events is:

$$\tfrac{1}{2}(-p \log p - (1 - p) \log (1 - p))$$

Accordingly, we see that, for the symmetrical binary channel, the equivocation, the sum of these terms, is

$$H_y(x) = -p \log p - (1 - p) \log (1 - p)$$

Thus the channel capacity C of the symmetrical binary channel is, from 8.8,

$$C = 1 + p \log p + (1 - p) \log (1 - p)$$

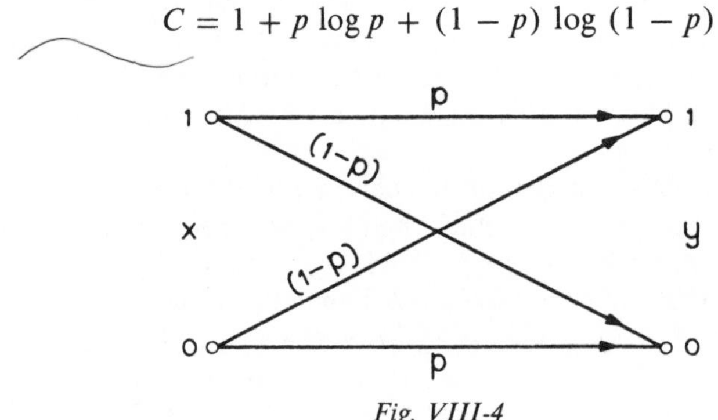

Fig. VIII-4

We should note that this channel capacity C is just unity less the function plotted against p in Figure V-1. We see that if p is ½, the channel capacity is 0. This is natural, for in this case, if we receive a 1, it is equally likely that a 1 or a 0 was transmitted, and the received message does nothing to resolve our uncertainty as to what digit the sender sent. We should also note that the channel capacity is the same for $p = 0$ as for $p = 1$. If we consistently receive a 0 when we transmit a 1 and a 1 when we transmit a 0, we are just as sure of the sender's intentions as if we always get a 1 for a 1 and a 0 for a 0.

If, on the average, 1 digit in 10 is in error, the channel capacity is reduced to .53 of its value for errorless transmission, and for one error in 100 digits, the channel capacity is reduced to .92 merely.

The writer would like to testify at this point that the simplicity of the result we have obtained for the symmetrical binary channel is in a sense misleading (it was misleading to the writer at least). The expression for the optimum rate (channel capacity) of an unsymmetrical binary channel in which the probability that a 1 is received as a 1 is p and the probability that a 0 is received as a 0 is a different number q is a mess, and more complicated channels must offer almost intractable problems.

Perhaps for this reason as well as for its practical importance, much consideration has been given to transmission over the symmetrical binary channel. What sort of codes are we to use in order to attain errorless transmission over such a channel? Examples devised by R. W. Hamming were mentioned by Shannon in his original paper. Later, Marcel J. E. Golay published concerning error-correcting codes in 1949, and Hamming published his work in 1950. We should note that these codes were devised subsequent to Shannon's work. They *might,* I suppose, have been devised before, but it was only when Shannon showed error-free transmission to be possible that people asked, "How can we achieve it?"

We have noted that to get an efficient correction of errors, the encoder must deal with a long sequence of message digits. As a simple example, suppose we encode our message digits in blocks of 16 and add after each block a sequence of *check digits* which enable us to detect a single error in *any one* of the digits, message digits or check digits. As a particular example, consider the sequence of message digits 1 1 0 1 0 0 1 1 0 1 0 1 1 0 0 0. To find the appropriate check digits, we write the 0's and 1's constituting the message digits in the 4 by 4 grid shown in Figure VIII-5. Associated with each row and each column is a circle. In each circle is a 0 or a 1 chosen so as to make the total number of 1's in the column or row (including the circle as well as the squares) even. Such added digits are called *check digits.* For the particular assortment of message digits used as an example, together with the appropriately chosen check digits, the numbers of 1's in successive columns (left to right) and 2, 2, 2, 4, all being even numbers, and the numbers of 1's in successive rows (top to bottom) are 4, 2, 2, 2, which are again all even.

What happens if a single error is made in the transmission of a message digit among the 16? There will be an odd number of ones *in a row and in a column.* This tells us to change the message digit where the row and column intersect.

What happens if a single error is made in a check digit? In this case there will be an odd number of ones *in a row or in a column.* We have detected an error, but we see that it was not among the message digits.

The total number of digits transmitted for 16 message digits is 16 + 8, or 24; we have increased the number of digits needed in the ratio 24/16, or 1.5. If we had started out with 400 message digits, we would have needed 40 check digits and we would have increased the number of digits needed only in the ratio of 440/400,

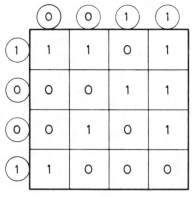

	(0)	(0)	(1)	(1)
(1)	1	1	0	1
(0)	0	0	1	1
(0)	0	1	0	1
(1)	1	0	0	0

Fig. VIII-5

or 1.1. Of course, we would have been able to correct only one error in 440 rather than one error in 24.

Codes can be devised which can be used to correct larger numbers of errors in a block of transmitted characters. Of course, more check digits are needed to correct more errors. A final code, however we may devise it, will consist of some set of 2^M blocks of 0's and 1's representing all of the blocks of digits M digits long which we wish to transmit. If the code were not error correcting, we could use a block just M digits long to represent each block of M digits which we wish to transmit. We will need more digits per block because of the error-correcting feature.

When we receive a given block of digits, we must be able to deduce from it which block was sent despite some number n of errors in transmission (changes of 0 to 1 or 1 to 0). A mathematician would say that this is possible if the distance between any two blocks of the code is at least $2n + 1$.

Here *distance* is used in a queer sense indeed, as defined by the mathematician for his particular purpose. In this sense, the distance between two sequences of binary digits is the number of 0's or 1's that must be changed in order to convert one sequence into the other. For instance, the distance between 0 0 1 0 and 1 1 1 1 is 3, because we can convert one sequence into the other only by changing three digits *in one sequence or in the other.*

When we make n errors in transmission, the block of digits we receive is a distance n from the code word we sent. It *may be* a distance n digits closer to some other code word. If we want to be sure that the received block will always be nearer to the correct code word, the one that was sent, than to any other code word, then the distance from any code word to any other code word must be at least $2n + 1$.

Thus, one problem of block coding is to find 2^M equal length code words (longer than M binary digits) that are all at least a distance $2n + 1$ from one another. The code words must be as short as possible. The codes of Hamming and Golay are efficient, and other efficient codes have been found.

Another problem of block coding is to provide a feasible scheme for encoding and, especially, for decoding. Simply listing code words won't do. The list would be too long. Encoding blocks of 20 binary digits ($M = 20$) requires around a million code words. And, finding the code word nearest to some received block of digits would take far too long.

Algebraic coding theory provides means for coding and decoding with the correction of many errors. Slepian was a pioneer in this field and important contributors can be identified by the names of types of algebraic codes: Reed-Solomon codes and Bose-Chaudhuri-Hocquenghem codes provide examples. Elwin Berlekamp contributed greatly to mathematical techniques for calculating the nearest code word more simply.

Convolutional codes are another means of error correction. In convolution coding, the latest *M* digits of the binary stream to be sent are stored in what is called a *shift register*. Every time a new binary digit comes in, 2 (or 3, or 4) are sent out by the coder. The digits sent out are produced by what is called modulo 2 addition of various digits stored in the shift register. (In modulo 2 addition of binary numbers one doesn't "carry.")

Convolutional encoding has been traced to early ideas of Elias, but the earliest coding and decoding scheme published is that in a patent of D. W. Hagelbarger, filed in 1958. Convolutional decoding really took off in 1967 when Andrew J. Viterbi invented an optimum and simple decoding scheme called maximum likelihood decoding.

Today, convolutional decoding is used in such valuable, noisy communication channels as in sending pictures of Jupiter and its satellites back from the Voyager spacecraft. Convolutional coding is particularly valuable in such applications because Viterbi's maximum likelihood decoding can take advantage of the *strength* as well as the *sign* of a received pulse.

If we receive a very small positive pulse, it is almost as likely to be a negative pulse plus noise as it is to be a positive pulse plus noise. But, if we receive a large positive pulse, it is much likelier to be a positive pulse plus noise than a negative pulse plus noise. Viterbi decoding can take advantage of this.

Block coding is used in protecting the computer storage of vital information. It can also be used in the transmission of binary information over inherently low-noise data circuits.

Many existing circuits that are used to transmit data are subject to long bursts of noise. When this is so, the most effective form of error correction is to divide the message up into long blocks of digits and to provide foolproof error detection. If an error is detected in a received block, retransmission of the block is requested.

Mathematicians are fascinated by the intricacies and challenges of block coding. In the eyes of some, information theory has become essentially algebraic coding theory. Coding theory is important to information theory. But, in its inception, in Shannon's work, information theory was, as we have seen, much broader. And even in coding itself, we must consider source coding as well as channel coding.

In Chapter VII, we discussed ways of removing redundancy from a message so that it could be transmitted by means of fewer binary digits. In this chapter, we have considered the matter of adding redundancy to a nonredundant message in order to attain virtually error-free transmission over a noisy channel. The fact that such error-free transmission *can* be attained using a noisy channel was and is surprising to communication engineers and mathematicians, but Shannon has proved that it is necessarily so.

Prior to receiving a message over an error-free channel, the recipient is uncertain as to what particular message out of many possible messages the sender will actually transmit. The amount of the recipient's uncertainty is the entropy or information rate of the message source, measured in bits per symbol or per second. The recipient's uncertainty as to what message the message source will send is completely resolved if he receives an exact replica of the message transmitted.

A message may be transmitted by means of positive and negative pulses of current. If a strong enough noise consisting of random positive and negative pulses is added to the signal, a positive signal pulse may be changed into a negative pulses or a negative signal pulse may be changed into a positive pulse. When such a noisy channel is used to transmit the message, if the sender sends any particular symbol there is some uncertainty as to what symbol will be received by the recipient of the message.

When the recipient receives a message over a noisy channel, he knows what message he has received, but he cannot ordinarily be sure what message was transmitted. Thus, his uncertainty as to what message the sender chose is not completely resolved even on the receipt of a message. The remaining uncertainty depends on the probability that a received symbol will be other than the symbol transmitted.

From the sender's point of view, the uncertainty of the recipient as to the true message is the uncertainty, or entropy, of the message source plus the uncertainty of the recipient as to what message was transmitted *when he knows what message was received.* The measure which Shannon provides of this latter uncertainty is the *equivocation,* and he defines the rate of transmission of information as the entropy of the message source less the equivocation.

The rate of transmission of information depends both on the amount of noise or uncertainty in the channel and on what message source is connected to the channel at the transmitting end. Let us suppose that we choose a message source such that this rate of transmission which we have defined is as great as it is possible to make it. This greatest possible rate of transmission is called the *channel capacity* for a noisy channel. The channel capacity is measured in bits per symbol or per second.

So far, the channel capacity is merely a mathematically defined quantity which we can compute if we know the probabilities of various sorts of errors in the transmission of symbols. The channel capacity is important, because Shannon proves, as his fundamental theorem for the noisy channel, that when the entropy or information rate of a message source is less than this channel capacity, the messages produced by the source can be so encoded that they can be transmitted over the noisy channel with an error less than any specified amount.

In order to encode messages for error-free transmission over noisy channels, long sequences of symbols must be lumped together and encoded as one supersymbol. This is the sort of block encoding that we have encountered earlier. Here we are using it for a new purpose. We are not using it to remove the redundancy of the messages produced by a message source. Instead, we are using it to add redundancy to nonredundant messages so that they can be transmitted without error over a noisy channel. Indeed, the whole problem of efficient and error-free communication turns out to be that of removing from messages the somewhat inefficient redundancy which they have and then adding redundancy of the right sort in order to allow correction of errors made in transmission.

The redundant digits we must use in encoding messages for error-free transmission, of course, slow the speed of transmission. We have seen that in using a binary symmetric channel in which 1 transmitted digit in 100 is erroneously received, we can send only 92 correct nonredundant message digits for each 100 digits we feed into the noisy channel. This means that on the average, we must use a redundant code in which, for each 92 nonredundant message digits, we must include in some way 8 extra check digits thus making the over-all stream of digits redundant.

Shannon's very general work tells us in principle how to proceed. But, the mathematical difficulties of treating complicated channels are great. Even in the case of the simple, symmetric, off-on binary channel, the problem of finding efficient codes is formidable, although mathematicians have found a large number of best codes. Alas, even these seem to be too complicated to use!

Is this a discouraging picture? How much wiser we are than in the days before information theory! We know what the problem is. We know in principle how well we can do, and the result has astonished engineers and mathematicians. Further, we do have effective error-correcting codes that are used in a variety of applications, including the transmission back to earth of glamorous pictures of far planets.

CHAPTER IX *Many Dimensions*

YEARS AND YEARS AGO (over thirty) I found in the public library of St. Paul a little book which introduced me to the mysteries of the fourth dimension. It was *Flatland,* by Abbott. It describes a two-dimensional world without thickness. Such a world and all its people could be drawn in complete detail, inside and out, on a sheet of paper.

What I now most remember and admire about the book are the descriptions of Flatland society. The inhabitants are polygonal, and sidedness determines social status. The most exalted of the multisided creatures hold the honorary status of circles. The lowest order is isosceles triangles. Equilateral triangles are a step higher, for regularity is admired and required. Indeed, irregular children are cracked and reset to attain regularity, an operation which is frequently fatal. Women are extremely narrow, needle-like creatures and are greatly admired for their swaying motion. The author of record, A. Square, accords well with all we have come to associate with the word.

Flatland has a mathematical moral as well. The protagonist is astonished when a circle of varying size suddenly appears in his world. The circle is, of course, the intersection of a three-dimensional creature, a sphere, with the plane of Flatland. The sphere explains the mysteries of three dimensions to A. Square, who in turn preaches the strange doctrine. The reader is left with the thought that he himself may someday encounter a fluctuating and disappearing entity, the three-dimensional intersection of a four-dimensional creature with our world.

Four-dimensional cubes or *tesseracts,* hyperspheres, and other hypergeometric forms are old stuff both to mathematicians and to science fiction writers. Supposing a fourth dimension like unto the three which we know, we can imagine many three-dimensional worlds existing as close to one another as the pages of a manuscript, each imprinted with different and distinct characters and each separate from every other. We can imagine traveling through the fourth dimension from one world to another or reaching through the fourth dimension into a safe to steal the bonds or into the abdomen to snatch an appendix.

Most of us have heard also that Einstein used time as a fourth dimension, and some may have heard of the many-dimensional *phase spaces* of physics, in which the three coordinates and three velocity components of each of many particles are all regarded as dimensions.

Clearly, this sort of thing is different from the classical idea of a fourth spatial dimension which is just like the three dimensions of up and down, back and forth, and left and right, those we all know so well. The truth of the matter is that nineteenth-century mathematicians succeeded in generalizing geometry to include any number of dimensions or even an infinity of dimensions.

These dimensions are for the pure mathematician merely mental constructs. He starts out with a line called the x direction or *x axis,* as shown in a of Figure IX-1. Some point p lies a distance x_p to the right of the *origin O* on the x axis. This *coordinate* x_p in fact describes the location of the point p.

The mathematician can then add a y axis perpendicular to the x axis, as shown in *b* of Figure IX-1. He can specify the location of a point p in the two-dimensional space or plane in which these axes lie by means of two numbers or coordinates: the distance from the origin O in the y direction, that is, the height y_p, and the distance from the origin O in the x direction x_p, that is, how far p is to the right of the origin O.

In *c* of Figure IX-1 the x, y, and z axes are supposed to be all perpendicular to one another, like the edges of a cube. These axes represent the directions of the three-dimensional space with which we are familiar. The location of the point p is given by its height y_p above the origin O, its distance x_p to the right of the origin O, and its distance z_p behind the origin O.

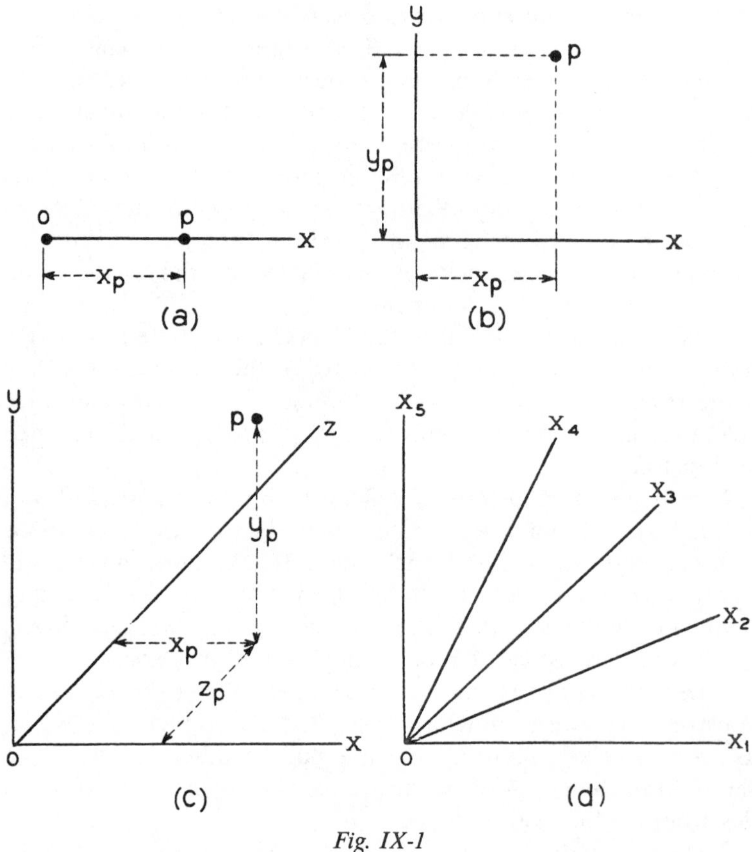

Fig. IX-1

Of course, in the drawing *c* of Figure IX-1 the *x, y,* and *z* axes aren't really all perpendicular to one another. We have here merely a two-dimensional perspective sketch of an actual three-dimensional situation in which the axes are all perpendicular to one another. In *d* of Figure IX-1, we similarly have a two-dimensional perspective sketch of axes in a five-dimensional space. Since we come to the end of the alphabet in going from *x* to *z,* we have merely labeled these directions x_1, x_2, x_3, x_4, x_5, according to the practice of mathematicians.

Of course these five axes of *d* of Figure IX-1 are not all perpendicular to one another in the drawing, but neither are the three axes of *c*. We can't lay out five mutually perpendicular lines in our three-dimensional space, but the mathematician can deal logically with a "space" in which five or more axes are mutually perpendicular. He can reason out the properties of various geometrical figures in a five-dimensional space, in which the position of a point p is described by five coordinates $x_{1p}, x_{2p}, x_{3p}, x_{4p}, x_{5p}$. To make the space like ordinary space (a *Euclidean* space) the mathematician says that the square of the distance d of the point p from the origin shall be given by

$$d^2 = x_{1p}{}^2 + x_{2p}{}^2 + x_{3p}{}^2 + x_{4p}{}^2 + x_{5p}{}^2 \tag{9.1}$$

In dealing with multidimensional spaces, mathematicians define the "volume" of a "cubical" figure as the product of the lengths of its sides. Thus, in a two-dimensional space the figure is a square, and, if the length of each side is L, the "volume" is the area of the square, which is L^2. In three-dimensional space the volume of a cube of width, height, and thickness L is L^3. In five-dimensional space the volume of a hypercube of extent L in each direction is L^5, and a ninety-nine dimensional cube L on a side would have a volume L^{99}.

Some of the properties of figures in multidimensional space are simple to understand and startling to consider. For instance, consider a circle of radius 1 and a concentric circle of radius ½ inside of it, as shown in Figure IX-2. The area ("volume") of a circle is πr^2, so the area of the outer circle is π and the area of the inner

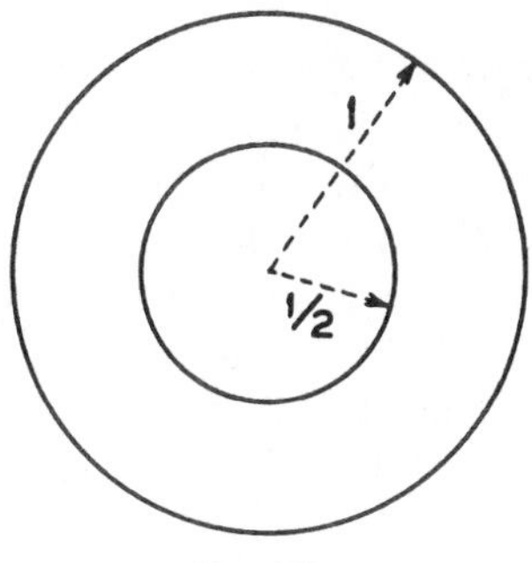

Fig. IX-2

circle is $\pi(1/2)^2 = (1/4)\pi$. Thus, a quarter of the area of the whole circle lies within a circle of half the diameter.

Suppose, however, that we regard Figure IX-2 as representing spheres. The volume of a sphere is $(4/3)\pi r^3$, and we find that 1/8 of the volume of a sphere lies within a sphere of 1/2 diameter. In a similar way, the volume of a hypersphere of n dimensions is proportional to r^n, and as a consequence the fraction of the volume which lies in a hypersphere of half the radius is $1/2^n$. For instance, for $n = 7$ this is a fraction 1/128.

We could go through a similar argument concerning the fraction of the volume of a hypersphere of radius r that lies within a sphere of radius $0.99r$. For a 1,000-dimension hypersphere we find that a fraction 0.00004 of the volume lies in a sphere of 0.99 the radius. The conclusion is inescapable that in the case of a hypersphere of a very high dimensionality, essentially all of the volume lies very near to the surface!

Are such ideas anything but pure mathematics of the most esoteric sort? They are pure and esoteric mathematics unless we attach them to some problem pertaining to the physical world. Imaginary numbers, such as $\sqrt{-1}$, once had no practical physical meaning. However, imaginary numbers have been assigned meanings in electrical engineering and physics. Can we perhaps find a physical situation which can be represented accurately by the mathematical properties of hyperspace? We certainly can, right in the field of communication theory. Shannon has used the geometry of multidimensional space to prove an important theorem concerning the transmission of continuous, band-limited signals in the presence of noise.

Shannon's work provides a wonderful example of the use of a new point of view and of an existing but hitherto unexploited branch of mathematics (in this case, the geometry of multidimensional spaces) in solving a problem of great practical interest. Because it seems to me so excellent an example of applied mathematics, I propose to go through a good deal of Shannon's reasoning. I believe that the course of this reasoning is more unfamiliar than difficult, but the reader will have to embark on it at his own peril.

In order to discuss this problem of transmission of continuous signals in the presence of noise, we must have some common measure of the strength of the signal and of the noise. *Power* turns out to be an appropriate and useful measure.

When we exert a force of 1 lb over a distance of 1 ft in raising a 1 lb weight to the height of 1 ft we do *work*. The amount of work done is 1 *foot-pound* (ft-lb). The weight has, by virtue of its height, an *energy* of 1 ft-lb. In falling, the weight can do an amount of work (as in driving a clock) equal to this energy.

Power is rate of doing work. A machine which expends 33,000 ft-lb of energy and does 33,000 ft-lb of work in a minute has by definition a power of 1 *horsepower* (hp).

In electrical calculations, we reckon energy and work in terms of a unit called the *joule* and power in terms of a unit called a *watt*. A watt is one joule per second.

If we double the voltage of a signal, we increase its energy and power by a factor of 4. Energy and power are proportional to the square of the voltage of a signal.

We have seen as far back as Chapter IV that a continuous signal of band width W can be represented completely by its amplitude at $2W$ sample points per second. Conversely, we can construct a band-limited signal which passes through any $2W$ sample points per second which we may choose. We can specify each sample arbitrarily and change it without changing any other sample. When we so change any sample we change the corresponding band-limited signal.

We can measure the amplitudes of the samples in volts. Each sample represents an energy proportional to the square of its voltage.

Thus, we can express the squares of the amplitudes of the samples in terms of energy. By using rather special units to measure energy, we can let the energy be equal to the square of the sample amplitude, and this won't lead to any troubles.

Let us, then, designate the amplitudes of successive and correctly chosen samples of a band-limited signal, measured perhaps in volts, by the letters x_1, x_2, x_3, etc. The parts of the signal energy represented by the samples will be x_1^2, x_2^2, x_3^2, etc. The total energy of the signal, which we shall call E, will be the sum of these energies:

$$E = x_1^2 + x_2^2 + x_3^2 + \text{etc.} \tag{9.2}$$

But we see that in geometrical terms E is just the square of the distance from the origin, as given by 9.1, if x_1, x_2, x_3, etc., are the coordinates of a point in multidimensional space!

Thus, if we let the amplitudes of the samples of a band-limited signal be the coordinates of a point in hyperspace, the point itself represents the complete signal, that is, all the samples taken together, and the square of the distance of the point from the origin represents the energy of the complete signal.

Why should we want to represent a signal in this geometrical fashion? The reason that Shannon did so was to prove an important theorem of communication theory concerning the effect of noise on signal transmission.

In order to see how this can be done, we should recall the mathematical model of a signal source which we adopted in Chapter III. We there assumed that the source is both stationary and ergodic. These assumptions must extend to the noise we consider and to the combined "source" of signal plus noise.

It is not actually impossible that such a source might produce a signal or a noise consisting of a very long succession of very high-energy samples or of very low-energy samples, any more than it is impossible that an ergodic source of letters might produce an extremely long run of E's. It is merely very unlikely. Here we are dealing with the theorem we encountered first in Chapter V. An ergodic source can produce a class of messages which are probable and a class which are so very improbable that we can disregard them. In this case, the improbable messages are those for which the average power of the samples produced departs significantly from the time average (and the ensemble average) characteristic of the ergodic source.

Thus, for all the long messages that we need to consider, there is a meaningful average power of the signal which does not change appreciably with time. We can measure this average power by adding the energies of a large number of successive samples and dividing by the time T during which the samples are sent. As we make the time T longer and longer and the number of samples larger and larger, we will get a more and more accurate value for the average power. Because the source is stationary, this average power will be the same no matter what succession of samples we use.

We can say this in a different way. Except in cases so unlikely that we need not consider them, the total energy of a large number of successive samples produced by a stationary source will be nearly the same (to a small fractional difference) regardless of what particular succession of samples we choose.

Because the signal source is ergodic as well as stationary, we can say more. For each signal the source produces, regardless of what the particular signal is, it is practically certain that the energy of the same large number of successive samples will be nearly the same, and the fractional differences among energies get smaller and smaller as the number of samples is made larger and larger.

Let us represent the signals from such a source by points in hyperspace. A signal of band width W and duration T can be represented by $2WT$ samples, and the amplitude of each of these samples is the distance along one coordinate axis of hyperspace. If the average energy per sample is P, the total energy of the $2WT$ samples will be very close to $2WTP$ if $2WT$ is a very large number of samples. We have seen that this total energy tells how far from the origin the point which represents the signal is. Thus, as the number of samples is made larger and larger, the points representing different signals of the same duration produced by the source lie within a smaller and smaller distance (measured as a fraction of the radius) from the surface of a hypersphere of radius $\sqrt{2WTP}$. The fact that the points representing the different signals all lie so close to the surface is not surprising if we remember that for a hypersphere of high dimensionality almost all of the volume is very close to the surface.

We receive, not the signal itself, but the signal with noise added. The noise which Shannon considers is called *white Gaussian noise.* The word *white* implies that the noise contains all frequencies equally, and we assume that the noise contains all frequencies equally up to a frequency of W cycles per second and no higher frequencies. The word *Gaussian* refers to a law for the probability of samples of various amplitudes, a law which holds for many natural sources of noise. For such Gaussian noise, each of the $2W$ samples per second which represent it is uncorrelated and independent. If we know the average energy of the samples which we will call N, knowing the energy of some samples doesn't help to predict the energy of others. The total energy of $2WT$ samples will be very nearly $2WTN$ if $2WT$ is a large number of samples, and the energy will be almost the same for any succession of noise samples that are added to the signal samples.

We have seen that a particular succession of signal samples is represented by some point in hyperspace a distance $\sqrt{2WTP}$ from the origin. The sum of a signal plus noise is represented by some point a little distance away from the point representing the signal alone. In fact, we see that the distance from the point representing the signal alone to the point representing the signal plus the noise is $\sqrt{2WTN}$. Thus, the signal plus the noise lies in a little hypersphere of radius $\sqrt{2WTN}$ centered on the point representing the signal alone.

Now, we don't receive the signal alone. We receive a signal of average energy P per sample plus Gaussian noise of average energy N per sample. In a time T, the total received energy is $2WT(P + N)$ and the point representing whatever signal was sent plus whatever noise was added to it lies within a hypersphere of radius $\sqrt{2WT(P + N)}$.

After we have received a signal plus noise for T seconds we can find the location of the point representing the signal plus noise. But how are we to find the signal? We only know that the signal lies within a distance $\sqrt{2WTN}$ of the point representing the signal plus noise.

How can we be sure of deducing what signal was sent? Suppose that we put into the hypersphere of radius $\sqrt{2WT(P + N)}$, in which points representing a signal plus noise must lie, a large number of little nonoverlapping hyperspheres of radius a bare shade larger than $\sqrt{2WTN}$. Let us then send only signals represented by the center points of these little spheres.

When we receive the $2WT$ samples of any particular one of these signals plus any noise samples, the corresponding point in hyperspace can only lie within the particular little hypersphere surrounding that signal point and not within any other. This is so because, as we have noted, the points representing long sequences of samples produced by an ergodic noise source must be almost at the surface of a sphere of radius $\sqrt{2WTN}$. Thus, the signal sent can be identified infallibly despite the presence of the noise.

How many such nonoverlapping hyperspheres of radius $\sqrt{2WTN}$ can be placed in a hypersphere of radius $\sqrt{2WT(P+N)}$? The number certainly cannot be larger than the ratio of the volume of the larger sphere to that of the smaller sphere.

The number n of dimensions in the space is equal to the number of signal (and noise) samples $2WT$. The volume of a hypersphere in a space of n dimensions is proportional to r^n. Hence, the ratio of the volume of the large signal-plus-noise sphere to the volume of the little noise sphere is

$$\left(\frac{\sqrt{2WT(P+N)}}{\sqrt{2WTN}}\right)^{2WT} = \left(\frac{P+N}{N}\right)^{WT}$$

This is a limit to the number of distinguishable messages we can transmit in a time T. The logarithm of this number is the number of bits which we can transmit in the time T. It is

$$WT \log \left(\frac{P+N}{N}\right)$$

As the message is T seconds long, the corresponding number of bits per second C is

$$C = W \log (1 + P/N) \tag{9.3}$$

Having got to this point, we can note that the ratio of average energy per signal sample to average energy per noise sample must be equal to the ratio of average signal power to average noise power, and we can, in 9.3, regard P/N as the ratio of signal power to noise power instead of as the ratio of average signal-sample energy to average noise-sample energy.

The foregoing argument, which led to 9.3, has merely shown that no more than C bits per second can be sent with a band width of W cycles per second using a signal of power P mixed with a Gaussian noise of power N. However, by a further geometrical argument, in which he makes use of the fact that the volume of a hypersphere of high dimensionality is almost all very close to the surface, Shannon shows that the signaling rate can approach as close as one likes to C as given by 9.3 with as small a number of errors as one likes. Hence, C, as given by 9.3, is the channel capacity for a continuous channel in which a Gaussian noise is added to the signal.

It is perhaps of some interest to compare equation 9.3 with the expressions for speed of transmission and for information which Nyquist and Hartley proposed in 1928 and which we discussed in Chapter II. Nyquist and Hartley's results both say that the number of binary digits which can be transmitted per second is

$$n \log m$$

Here m is the number of different symbols, and n is the number of symbols which are transmitted per second.

One sort of symbol we can consider is a particular value of voltage, as, $+3$, $+1$, -1, or -3. Nyquist knew, as we do, that the number of independent samples or values of voltage which can be transmitted per second is $2W$. By using this fact, we can rewrite equation 9.3 in the form

$$C = (n/2) \log (1 + P/N)$$
$$C = n \log \sqrt{(1 + P/N)}$$

Here we are really merely retracing the steps which led us to 9.3.

We see that in equation 9.3 we have got at the average number m of different symbols we can send per sample, in terms of the ratio of signal power to noise power. If the signal power becomes very small or the noise power becomes very large, so that P/N is nearly 0, then the average number of different symbols we can transmit per sample goes to

$$\log 1 = 0$$

Thus, the average number of symbols we can transmit per sample and the channel capacity go to 0 as the ratio of signal power to noise power goes to 0. Of course, the number of symbols we can transmit per sample and the channel capacity become large as we make the ratio of signal power to noise power large.

Our understanding of *how* to send a large average number of independent symbols per sample has, however, gone far beyond anything which Nyquist or Hartley told us. We know that if we are to do this most efficiently we must, in general, not try to encode a symbol for transmission as a particular sample voltage to be sent by itself. Instead, we must, in general, resort to the now-familiar procedure of block encoding and encode a long sequence of symbols by means of a large number of successive samples. Thus, if the ratio of signal power to noise power is 24, we can on the average transmit with negligible error $\sqrt{1 + 24} = \sqrt{25} = 5$ different symbols per sample, but we can't transmit any of 5 different symbols by means of one particular sample.

In Figure VIII-1 of Chapter VIII, we considered sending binary digits one at a time in the presence of noise by using a signal which was either a positive or a negative pulse of a particular amplitude and calling the received signal a 1 if the signal plus noise was positive and a 0 if the received signal plus noise was negative. Suppose that in this case we make the signal powerful enough compared with the noise, which we assume to be Gaussian, so that only 1 received digit in 100,000 will be in error. Calculations show that this calls for about six times the signal power which equation 9.3 says we will need for the same band width and noise power. The extra power is needed because we use as a signal either a short positive or negative pulse specifying one binary digit, rather than using one of many long signals consisting of many different samples of various amplitudes to represent many successive binary digits.

One very special way of approaching the ideal signaling rate or channel capacity for a small, average signal power in a large noise power is to concentrate the signal power in a single short but powerful pulse and to send this pulse in one of many possible time positions, each of which represents a different symbol. In this very special and unusual case we *can* efficiently transmit symbols one at a time.

If we wish to approach Shannon's limit for a chosen bandwidth we must use as elements of the code long, complicated signal waves that resemble gaussian noise.

We can if we wish look on relation 9.3 not narrowly as telling us how many bits per second we can send over a particular communication channel but, rather, as telling us something about the possibilities of transmitting a signal of a specified band width with some required signal-to-noise ratio over a communication channel of some other band width and signal-to-noise ratio. For instance, suppose we must send a signal with a band width of 4 megacycles per second and attain a ratio of signal power to noise power P/N of 1,000. Relation 9.3 tells us that the corresponding channel capacity is

$$C = 40{,}000{,}000 \text{ bits/second}$$

But the same channel capacity can be attained with the combinations shown in Table XIII.

TABLE XIII

Combinations of W and P/N Which Give Same Channel Capacity

W	P/N
4,000,000	1,000
8,000,000	30.6
2,000,000	1,000,000

We see from Table XIII that, in attaining a given channel capacity, we can use a broader band width and a lower ratio of signal to noise or a narrower band width and a higher ratio of signal to noise.

Early workers in the field of information theory were intrigued with the idea of cutting down the band width required by increasing the power used. This calls for lots of power. Experience has shown that it is much more useful and practical to increase the band width so as to get a good signal-to-noise ratio with less power than would otherwise be required.

This is just what is done in FM transmission, as an example. In FM transmission, a particular amplitude of the message signal to be transmitted, which may, for instance, be music, is encoded as a radio signal of a particular frequency. As the amplitude of the message signal rises and falls, the frequency of the FM signal which represents it changes greatly, so that in sending high fidelity music which has a band width of 15,000 cps, the FM radio signal can range over a band width of 150,000 cps. Because FM transmission makes use of a band width much larger than that of the music of which it is an encoding, the signal-to-noise ratio of the received music can be much higher than the ratio of signal power to noise power in the FM signal that the radio receiver receives. FM is not, however, an ideally efficient system; it does not work the improvement which we might expect from 9.3.

Ingenious inventors are ever devising improved systems of modulation. Twice in my experience someone has proposed to me a system which purported to do better than equation 9.3, for the ideal channel capacity, allows. The suggestions were plausible, but I knew, just as in the case of perpetual motion machines, that something had to be wrong with them. Careful analysis showed where the error lay. Thus, communication theory can be valuable in telling us what can't be accomplished as well as in suggesting what can be.

One thing that can't be accomplished in improving the signal-to-noise ratio by increasing the band width is to make a system which will behave in an orderly and happy way for all ratios of signal power to noise power.

According to the view put forward in this chapter, we look on a signal as a point in a multidimensional space, where the number of dimensions is equal to the number of samples. To send a narrow-band signal of a few samples by means of a broad-band signal having more samples, we must in some way map points in a space of few dimensions into points in a space of more dimensions in a one-to-one fashion.

Way back in Chapter I, we proved a theorem concerning the mapping of points of a space of two dimensions (a plane) onto points of a space of one dimension (a line). We proved that if we map each point of the plane in a one-to-one fashion into a single corresponding point on the line, the mapping cannot be continuous. That is, if we move smoothly along a path in the plane from point to nearby point, the corresponding positions on the line *must* jump back and forth discontinuously. A similar theorem is true for the mapping of the points of any space onto a space of different dimensionality. This bodes trouble for transmission schemes in which few message samples are represented by many signal samples.

Shannon gives a simple example of this sort of trouble, which is illustrated in Figure IX-3. Suppose that we use two sample amplitudes v_2 and v_1 to represent a single sample amplitude u. We regard v_2 and v_1 as the distance up from and to the right of the lower left hand corner of a square. In the square, we draw a snaky line which starts near the lower left-hand corner and goes back and forth across the square, gradually progressing upward. We let distance along this line, measured from its origin near the lower left-hand corner to some specified point along the line, be u, the voltage or amplitude of the signal to be transmitted.

Certainly, any value of u is represented by particular values of v_1 and v_2. We see that the range of v_1 or v_2 is less than the range of u. We can transmit v_1 and v_2 and then reconstruct u with great accuracy. Or can we?

Suppose a little noise gets into v_1 and v_2, so that, when we try to find the corresponding value of u at the receiver, we land somewhere in a circle of uncertainty due to noise. As long as the diameter of the circle is less than the distance between the loops of the snaky path, we can tell what the correct value of u is to a fractional error much smaller than the fractional error of v_1 or v_2.

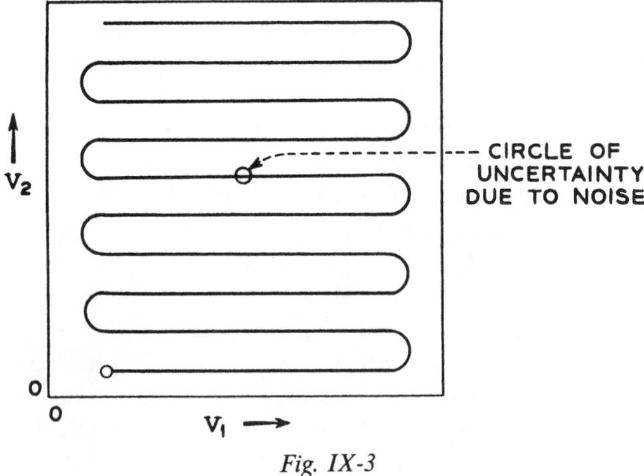

Fig. IX-3

But if the noise is larger, we can't be sure which loop of the snaky path was intended, and we frequently make a larger error in *u*.

This sort of behavior is inevitable in systems, such as FM, which use a large band width in order to get a better signal-to-noise ratio. As the noise added in transmission is increased, the noise in the received (demodulated) signal at first increases gradually and then increases catastrophically. The system is said to "break" at this level of signal to noise. Here we have an instance in which a seemingly abstract theorem of mathematics tells us that a certain type of behavior cannot be avoided in electrical communication systems of a certain general type.

The approach in this chapter has been essentially geometrical. This is only one way of dealing with the problems of continuous signals. Indeed, Shannon gives another in his book on communication theory, an approach which is applicable to all types of signals and noise. The geometrical approach is interesting, however, because it is proving illuminating and fruitful in many problems concerning electric signals which are not directly related to communication theory.

Here we have arrived at a geometry of band-limited signals by sampling the signals and then letting the amplitudes of the samples be the coordinates of a point in a multidimensional space. It is possible, however, to geometrize band-limited signals without speaking in terms of samples, and mathematicians interested in problems of signal transmission have done this. In fact, it is becoming increasingly common to represent band-limited signals as points in a multidimensional "signal space" or "function space" and to prove theorems about signals by the methods of geometry.

The idea of signals as points in a multidimensional signal space or function space is important, because it enables mathematicians to think about and to make statements which are true about *all* band-limited signals, or about large classes of band-limited signals, without considering the confusing details of particular signals, just as mathematicians can make statements about *all* triangles or *all right triangles.* Signal space is a powerful tool in the hands or, rather, in the minds of competent mathematicians. We can only wonder and admire.

From the point of view of communication theory, our chief concern in this chapter has been to prove an important theorem concerning a noisy continuous channel. The result is embodied in equation 9.3, which gives the rate at which we can transmit binary digits with negligible error over a continuous channel in which a signal of band width W and power P is mixed with a white Gaussian noise of band width W and power N.

Nyquist knew, in 1928, that one can send $2W$ independent symbols per second over a channel of band width $2W$, but he didn't know how many different symbols could be sent per second for a given ratio of signal power to noise power. We have found this out for the case of a particular, common type of noise. We also know that even if we can transmit some average number m of symbols per sample, in general, we can't do this by trying to encode successive symbols independently as particular voltages. Instead, we must use block encoding, and encode a large number of successive symbols together.

Equation 9.3 shows that we can use a signal of large band width and low ratio of signal power to noise power in transmitting a message which has a small band width and a large ratio of signal power to noise power. FM is an example of this. Such considerations will be pursued further in Chapter X.

This chapter has had another aspect. In it we have illustrated the use of a novel viewpoint and the application of a powerful field of mathematics in attacking a problem of communication theory. Equation 9.3 was arrived at by the by-no-means-obvious expedient of representing long electrical signals and the noises added to them by points in a multidimensional space. The square of the distance of a point from the origin was interpreted as the energy of the signal represented by the point.

Thus, a problem in communication theory was made to correspond to a problem in geometry, and the desired result was arrived at by geometrical arguments. We noted that the geometrical representation of signals has become a powerful mathematical tool in studying the transmission and properties of signals.

The geometrization of signal problems is of interest in itself, but it is also of interest as an example of the value of seeking new mathematical tools in attacking the problems raised by our increasingly complex technology. It is only by applying this order of thought that we can hope to deal with the increasingly difficult problems of engineering.

CHAPTER X *Information Theory and Physics*

I HAVE GIVEN SOMETHING of the historical background of communication theory in Chapter II. From this we can see that communication theory is an outgrowth of electrical communication, and we know that the behavior of electric currents and electric and magnetic fields is a part of physics.

To Morse and to other early telegraphists, electricity provided a very limited means of communication compared with the human voice or the pen in hand. These men had to devise codes by means of which the letters of the alphabet could be represented by turning an electric current successively on and off. This same problem of the representation of material to be communicated by various sorts of electrical signals has led to the very general ideas concerning encoding which are so important in communication theory. In this relation of encoding to particular physical phenomena, we see one link between communication theory and physics.

We have also noted that when we transmit signals by means of wire or radio, we receive them inevitably admixed with a certain amount of interfering disturbances which we call noise. To some degree, we can avoid such noise. The noise which is generated in our receiving apparatus we can reduce by careful design and by ingenious invention. In receiving radio signals, we can use an antenna which receives signals most effectively from the direction of the transmitter and which is less sensitive to signals coming from other directions. Further, we can make sure that our receiver responds only to the frequencies we mean to use and rejects interfering signals and noise of other frequencies.

Still, when all this is done, some noise will inevitably remain, mixed with the signals that we receive. Some of this noise may come from the ignition systems of automobiles. Far away from man-made sources, some may come from lightning flashes. But even if lightning were abolished, some noise would persist, as surely as there is heat in the universe.

Many years ago an English biologist named Brown saw small pollen particles, suspended in a liquid, dance about erratically in the field of his microscope. The particles moved sometimes this way and sometimes that, sometimes swiftly and sometimes slowly. This we call *Brownian motion.* Brownian motion is caused by the impact on the particles of surrounding molecules, which themselves execute even a wilder dance. One of Einstein's first major works was a mathematical analysis of Brownian motion.

The pollen grains which Brown observed would have remained at rest had the molecules about them been at rest, but molecules are always in random motion. It is this motion which constitutes heat. In a gas, a molecule moves in a disorganized way. It moves swiftly or slowly in straight lines between frequent collisions. In a liquid, the molecules jostle about in close proximity to one another but continually changing place, sometimes moving swiftly and sometimes slowly. In a solid, the molecules vibrate about their mean positions, sometimes with a large amplitude and sometimes with a small amplitude, but never moving much with respect to their nearest neighbors. Always, however, in gas, liquid, or solid, the molecules move, with an average energy due to heat which is proportional to the temperature above absolute zero, however erratically the speed and energy may vary from time to time and from molecule to molecule.

Energy of mechanical motion is not the only energy in our universe. The electromagnetic waves of radio and light also have energy. Electromagnetic waves are generated by changing currents of electricty. Atoms are positively charged nuclei surrounded by negative electrons, and molecules are made up of atoms. When the molecules of a substance vibrate with the energy of heat, relative motions of the charges in them can generate electromagnetic waves, and these waves have frequencies which include those of what we call radio, heat, and light waves. A hot body is said to *radiate* electromagnetic waves, and the electromagnetic waves that it emits are called *radiation.*

The rate at which a body which is held at a given temperature radiates radio, heat, and light waves is not the same for all substances. Dark substances emit more radiation than shiny substances. Thus, silver, which is called shiny because it reflects most of any waves of radio, heat, or light falling on it, is a poor radiator, while the carbon particles of black ink constitute a good radiator.

When radiation falls on a substance, the fraction that is reflected rather than absorbed is different for radiation of different frequencies, such as radio waves and light waves. There is a very general rule, however, that for radiation of a given frequency, the amount of radiation a substance emits at a given temperature is directly proportional to the fraction of any radiation falling on it which is absorbed rather than reflected. It is as if there were a skin around each substance which allowed a certain fraction of any radiation falling on it to pass through and reflected the rest, and as if the fraction that passed through the skin were the same for radiation either entering or leaving the substance.

If this were not so, we might expect a curious and unnatural (as we know the laws of nature) phenomenon. Let us imagine a completely closed box or furnace held at a constant temperature. Let us imagine that we suspend two bodies inside the furnace. Suppose (contrary to fact) that the first of these bodies reflected radiation well, absorbing little, and that it also emitted radiation strongly, while the second absorbed radiation well, reflecting little, but emitted radiation poorly. Suppose that both bodies started out at the same temperature. The first would absorb less radiation and emit more radiation than the second, while the second would absorb more radiation and emit less radiation than the first. If this were so, the second body would become hotter than the first.

This is not the case, however; all bodies in a closed box or furnace whose walls are held at a constant, uniform temperature attain just exactly the same temperature as the walls of the furnace, whether the bodies are shiny, reflecting little radiation and absorbing much, or whether they are dark, reflecting little radiation and absorbing much. This can be so only if the ability to absorb rather than reflect radiation and the ability to emit radiation go hand in hand, as they always do in nature.

Not only do all bodies inside such a closed furnace attain the same temperature as the furnace; there is also a characteristic intensity of radiation in such an enclosure. Imagine a part of the radiation inside the enclosure to strike one of the walls. Some will be reflected back to remain radiation in the enclosure. Some will be absorbed by the walls. In turn, the walls will emit some radiation, which will be added to that reflected away from the walls. Thus, there is a continual interchange of radiation between the interior of the enclosure and the walls.

If the radiation in the interior were very weak, the walls would emit more radiation than the radiation which struck and was absorbed by them. If the radiation in the interior were very strong, the walls would receive and absorb more radiation than they emitted. When the electromagnetic radiation lost to the walls is just equal to that supplied by the walls, the radiation is said to be in *equilibrium* with its material surroundings. It has an energy which increases with temperature, just as the energy of motion of the molecules of a gas, a liquid, or a solid increases with temperature.

The intensity of radiation in an enclosure does not depend on how absorbing or reflecting the walls of the enclosure are; it depends only on the temperature of the walls. If this were not so and we made a little hole joining the interior of a shiny, reflecting enclosure with the interior of a dull, absorbing enclosure at the same temperature, there would have to be a net flow of radiation through the hole from one enclosure to another at the same temperature. This never happens.

We thus see that there is a particular intensity of electromagnetic radiation, such as light, heat, and radio waves, which is characteristic of a particular temperature. Now, while eletromagnetic waves travel through vacuum, air, or insulating substances such as glass, they can be guided by wires. Indeed, we can think of the signal sent along a pair of telephone wires either in terms of the voltage between the wires and the current of electrons which flows in the wires, or in terms of a wave made up of electric and magnetic fields between and around the wires, a wave which moves along with the current. As we can identify electrical signals on wires with electromagnetic waves, and as hot bodies radiate electromagnetic waves, we should expect heat to generate some sort of electrical signals. J. B. Johnson, who discovered the electrical fluctuations caused by heat, described them, not in terms of electromagnetic waves but in terms of a fluctuating voltage produced across a resistor.

Once Johnson had found and measured these fluctuations, another physicist was able to find a correct theoretical expression for their magnitude by applying the principles of statistical mechanics. This second physicist was none other than H. Nyquist, who, as we saw in Chapter II, also contributed substantially to the early foundations of information theory.

Nyquist's expression for what is now called either *Johnson noise* or *thermal noise* is

$$\overline{V^2} = 4\,\mathrm{k}TRW \qquad (10.1)$$

Here $\overline{V^2}$ is the mean square noise voltage, that is, the average value of the square of the noise voltage, across the resistor. k is Boltzmann's constant:

$$\mathrm{k} = 1.37 \times 10^{-23} \text{ joule/degree}$$

T is the temperature of the resistor in degrees Kelvin, which is the number of Celsius or centigrade degrees (which are 5/9 as large as Fahrenheit degrees) above absolute zero. Absolute zero is $-273°$ centigrade or $-459°$ Fahrenheit. R is the resistance of the resistor measured in ohms. W is the band width of the noise in cycles per second.

Obviously, the band width W depends only on the properties of our measuring device. If we amplify the noise with a broad-band amplifier we get more noise than if we use a narrow-band amplifier of the same gain. Hence, we would expect more noise in a television receiver, which amplifies signals over a band width of several million cycles per second, than in a radio receiver, which amplifies signals having a band width of several thousand cycles per second.

We have seen that a hot resistor produces a noise voltage. If we connect another resistor to the hot resistor, electric power will flow to this second resistor. If the second resistor is cold, the power will heat it. Thus, a hot resistor is a potential source of noise power. What is the most noise power N that it can supply? The power is

$$N = \mathrm{k}TW \tag{10.2}$$

In some ways, 10.2 is more satisfactory than 10.1. For one thing, it has fewer terms; the resistance R no longer appears. For another thing, its form is suitable for application to somewhat different situations.

For instance, suppose that we have a radio telescope, a big parabolic reflector which focuses radio waves into a sensitive radio receiver. I have indicated such a radio telescope in Figure X-1. Suppose we point the radio telescope at different celestial or terrestrial objects, so as to receive the electromagnetic noise which they radiate because of their temperature.

We find that the radio noise power received is given by 10.2, where T is the temperature of the object at which the radio telescope points.

If we point the telescope down at water or at smooth ground, what it actually sees is the reflection of the sky, but if we point it at things which don't reflect radio waves well, such as leafy trees or bushes, we get a noise corresponding to a temperature around 290° Kelvin (about 62° Fahrenheit), the temperature of the trees.

If we point the radio telescope at the moon and if the telescope

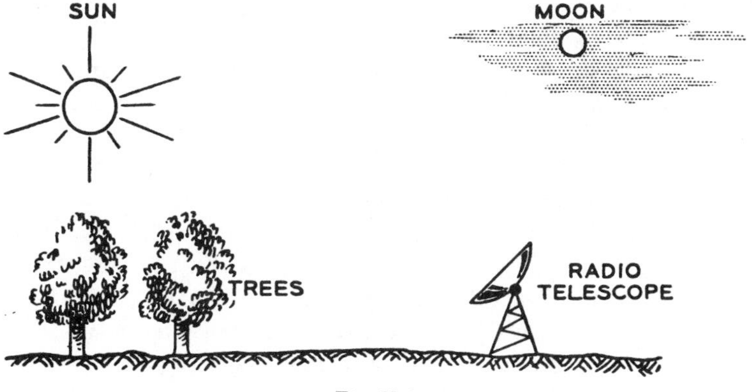

Fig. X-1

is directive enough to see just the moon and not the sky around it, we get about the same noise, which corresponds not to the temperature of the very surface of the moon but to the temperature a fraction of an inch down, for the substance of the moon is somewhat transparent to radio waves.

If we point the telescope at the sun, the amount of noise we obtain depends on the frequency to which we tune the radio receiver. If we tune the receiver to a frequency around 10 million cycles per second (a wave length of 30 meters), we get noise corresponding to a temperature of around a million degrees Kelvin; this is the temperature of the tenuous outer corona of the sun. The corona is transparent to radio waves of shorter wave lengths, just as the air of the earth is. Thus, if we tune the radio receiver to a frequency of around 10 billion cycles per second, we receive radiation corresponding to the temperature of around 8,000° Kelvin, the temperature a little above the visible surface. Just why the corona is so much hotter than the visible surface which lies below it is not known.

The radio noise from the sky is also different at different frequencies. At frequencies above a few billion cycles per second the noise corresponds to a temperature of about 3.5° Kelvin. At lower frequencies the noise is greater and increases steadily as the frequency is lowered. The Milky Way, particular stars, and island universes or galaxies in collision all emit large amounts of radio noise. The heavens are telling the age of the universe through their 3.5 degree microwave radiation, but other signals come from other places at other frequencies.

Nonetheless, Johnson or thermal noise constitutes a minimum noise which we must accept, and additional noise sources only make the situation worse. The fundamental nature of Johnson noise has led to its being used as a standard in the measurement of the performance of radio receivers.

As we have noted, a radio receiver adds a certain noise to the signals it receives. It also amplifies any noise that it receives. We can ask, how much amplified Johnson noise would just equal the noise the receiver adds? We can specify this noise by means of an equivalent noise temperature T_n. This equivalent noise temperature T_n is a measure of the noisiness of the radio receiver. The smaller T_n is the better the receiver is.

We can interpret the noise temperature T_n in the following way. If we had an ideal noiseless receiver with just the same gain and band width as the actual receiver and if we added Johnson noise corresponding to the temperature T_n to the signal it received, then the ratio of signal power to noise power would be the same for the ideal receiver with the Johnson noise added to the signal as for the actual receiver.

Thus, the noise temperature T_n is a just measure of the noisiness of the receiver. Sometimes another measure based on T_n is used; this is called the noise figure NF. In terms of T_n, the noise figure is

$$NF = \frac{293 + T_n}{293}$$

$$NF = 1 + \frac{T_n}{293} \tag{10.3}$$

The noise figure was defined for use here on earth, where every signal has mixed with it noise corresponding to a temperature of around 293° Kelvin. The noise figure is the ratio of the total output noise, including noise due to Johnson noise for a temperature of 293° Kelvin at the input and noise produced in the receiver, to the amplified Johnson noise alone.

Of course, the equivalent noise temperature T_n of a radio receiver depends on the nature and perfection of the radio receiver, and the lowest attainable noise figure depends on the frequency of operation. However, Table XIV below gives some rough figures for various sorts of receivers.

The effective temperatures of radio receivers and the tempera-

TABLE XIV

Type of Receiver	*Equivalent Noise Temperature, Degrees Kelvin*
Good Radio or TV receiver	1500°
Maser receiving station for space missions	20°
Parametric amplifier receiver	50°

tures of the objects at which their antennas are directed are very important in connection with communication theory, because noise determines the power required to send messages. Johnson noise is Gaussian noise, to which equation 9.3 applies. Thus, ideally, in order to transmit C bits per second, we must have a signal power P related to the noise power N by a relation that was derived in the preceding chapter:

$$C = W \log \left(1 + \frac{P}{N}\right) \qquad (9.3)$$

If we use expression 10.2 for noise, this becomes

$$C = W \log \left(1 + \frac{P}{kTW}\right) \qquad (10.4)$$

Let us assume a given signal power P. If we make W very small, C will become very small. However, if we make W larger and larger, C does not become larger and larger without limit, but rather it approaches a limiting value. When P/kTW becomes very small compared with unity, 10.4 becomes

$$C = \frac{1.44\,P}{kT} \qquad (10.5)$$

We can also write this

$$P = 0.693\,kTC \qquad (10.6)$$

Relation 10.6 says that, even when we use a very wide band width, we need at least a power 0.693 kT joule per second to send one bit per second, so that on the average we must use an energy of 0.693 kT joule for each bit of information we transmit. We should remember, however, that equation 9.3 holds only for an ideal sort of encoding in which many characters representing many bits of information are encoded together into a long stretch of signal. Most practical communication systems require much more energy per bit, as we noted in Chapter IX.

But, haven't we forgotten something? What about quantum effects? These may not be important in radio, but they are certainly important in communication by light. And, light has found a wide range of uses. Tiny optical fibers carry voice and other communication traffic. Pulses of light reflected from "corner reflectors" left by astronauts on the moon make it possible to track the earth-moon distance with an accuracy of about a tenth of a meter.

Harry Nyquist was a very forward-looking man. The derivation he gave for an expression for Johnson noise applies to all frequencies, including those of light. His expression for Johnson noise power in a bandwidth W was

$$N = \frac{\mathrm{h}fW}{e^{\mathrm{h}f/\mathrm{k}T} - 1} \qquad (10.7)$$

Here f is frequency in cycles per second and

$$\mathrm{h} = 6.63 \times 10^{-34} \text{ joule/sec.}$$

is Planck's constant. We commonly associate Planck's constant with the energy of a single photon of light, an energy E which is

$$E = \mathrm{h}f \qquad (10.8)$$

Quantum effects become important when hf is comparable to or larger than kT. Thus, the frequency above which the exact expression for Johnson noise, 10.7, departs seriously from the expression valid at low frequencies, 10.2, will be about

$$f = \mathrm{k}T/\mathrm{h} = 2.07 \times 10^{10}T \qquad (10.9)$$

When we take quantum effects into account, we find *less* not *more* noise at higher frequencies, and very much less noise at the frequency of light. But, there are quantum limitations other than those imposed by Johnson noise. It turns out that, ideally, 0.693 kT joules per bit is still the limit, even at the frequency of light. Practically, it is impossible to approach this limit closely. And, there is a common but wrong way of communicating which is very much worse. This is to amplify a weak received signal with the best sort of amplifier that is theoretically possible. Why is this bad?

At low frequencies, when we amplify a weak pulse we simply

obtain a pulse of greater power, and we can measure both the time that this stronger pulse peaks and the frequency spectrum of the pulse. But, the quantum mechanical uncertainty principle says that we cannot measure both time and frequency with perfect accuracy. In fact, if Δt is the error in measurement of time and Δf is the error in measurement of frequency, the best we can possibly do is

$$\Delta t \Delta f = 1$$
$$\Delta t = 1/\Delta f \tag{10.10}$$

The implication of 10.10 is that if we define the frequency of a pulse very precisely by making Δf small, we cannot measure the time of arrival of the pulse very accurately. In more mundane terms, we can't measure the time of arrival of a long narrow-band pulse as accurately as we can measure the time of arrival of a short broad-band pulse. But, just how are we frustrated in trying to make such a measurement?

Suppose that we do amplify a weak pulse with the best possible amplifier, and also shift all frequencies of the pulse down to some low range of frequencies at which quantum effects are negligible. We find that, mixed with the amplified signal, there will be a noise power N

$$N = G\,\mathrm{h}f\,W \tag{10.11}$$

Here f is the frequency of the original high-frequency signal, G is the power gain of the amplifying and frequency shifting system and W is the bandwidth. This noise is just enough to assure that we don't make measurements more accurately than allowed by the uncertainty principle as expressed in 10.10.

In order to increase the accuracy of a time measurement, we must increase the bandwidth W of the pulse. But, the added noise due to increasing the bandwidth, as expressed by 10.11, just undoes any gain in accuracy of measurement of time.

From 10.11 we can make the same argument that we made on page 192, and conclude that because of quantum effects we must use an energy of at least 0.693 hf joules per bit in order to communicate by means of a signal of frequency f. This argument is valid only for communication systems in which we amplify the received signal with the best possible amplifier—an amplifier which adds to the signal only enough noise to keep us from violating the uncertainty principle.

Is there an alternative to amplifying the weak received signal? At optical frequencies there certainly is. Quanta of light can be used to produce tiny pulses of electricity. Devices called photodiodes and photomultipliers produce a short electrical output pulse when a quantum of light reaches them—although they randomly fail to respond to some quanta. The *quantum efficiency* of actual devices is less than 100%.

Nonetheless, ideally we can detect the time of arrival of any quantum of light by using that quantum to produce a tiny electrical pulse. Isn't this contrary to the uncertainty principle? No, because when we measure the time of arrival of a quantum in this way we can't tell anything at all about the frequency of the quantum.

Photo detectors, commonly called *photon counters*, are used to measure the time of arrival of pulses of light reflected from the corner reflectors that astronauts left on the moon. They are also used in lightwave communication via optical fibers. Of course they aren't used as effectively as is ideally possible. What is the ideal limit? It turns out to be the same old limit we found on page 192, that is, 0.693 kT joules per bit. Quantum effects don't change that limiting performance, but they do make it far more difficult to attain. Why is this so?

The energy per photon is hf. Ideally, the energy per bit is 0.693 kT. Thus, ideally the bits per photon must be

$$\mathrm{h}f/(0.693\ \mathrm{k}T)$$

or

$$1.44\ (\mathrm{h}f/\mathrm{k}T) \text{ bits per photon}$$

How can we transmit many bits of energy by measuring the time of arrival of just a few photons, or one photon? We can proceed in this way. At the transmitter, we send out a pulse of light in one of M sub-intervals of some long time interval T. At the receiving end, the particular time interval in which we receive a photon conveys the message.

At best, this will make it possible to convey log M bits of information for each interval T. But, sometimes we will receive no photons in any time interval. And, sometimes thermal photons—Johnson noise photons—will arrive in a wrong time interval. This is what limits the transmission to 1.44 (hf/kT) bits per photon.

In reality, we can send far fewer bits per photon because it is impractical to implement effective systems that will send very many bits per photon.

We have seen through a combination of information theory and physics that it takes at least 0.693 kT joules of energy to convey one bit of information.

In most current radio receivers the noise actually present is greater than the environmental noise because the amplifier adds noise corresponding to some temperature higher than that of the environment. Let us use as the noise temperature T_n, the equivalent noise temperature corresponding to the noise actually added to the signal. How does actual performance compare with the ideal expression by equation (10.4)?

When we do not use error correction, but merely use enough signal power so that errors in receiving bits of information are very infrequent (around one error in 10^8 bits received), at best we must use about 10 times as much power per bit as calculated from expression 10.4.

The most sophisticated communication systems are those used in sending data back from deep-space missions. These are extremely low-noise maser receivers, and they make use of sophisticated error correction, including convolutional coding and decoding using the Viterbi algorithm. In sending pictures of Jupiter and its satellites back to earth, the Voyager spacecraft could transmit 115,200 binary digits per second with an error rate of one in 200 by using only 21.3 watts of power. The power is only 4.4 db more than the ideal limit using infinite bandwidth.

Pluto is about 6×10^{12} meters from earth. Ideally, how fast could we send data back from Pluto? We'll assume noise from space only, and no atmosphere absorption.

If we use a transmitting antenna of effective area A_T and a receiving antenna of effective area A_R, the ratio of received power

P_R to transmitted power P_T is given by Friis's transmission formula

$$\frac{P_R}{P_T} = \frac{A_R A_T}{\lambda^2 L^2} \tag{10.12}$$

where λ is the wavelength used in communication and L is the distance from transmitter to receiver, which in our case is 6×10^{12} meters.

Purely arbitrarily, we will assume a transmitter power of 10 watts. We will consider two cases. In the first, we use a wavelength of 1 cm or .01 meters. At this wavelength the temperature of space is indubitably 3.5° K, and we will assume that the transmitting antenna is 3.16 meters square, with an area of 10 square meters, and the receiving antenna is 31.6 meters square, with an area of 1,000 square meters. According to expression 10.12, if the transmitted power is 10 watts the received power will be 2.8×10^{-17} watts or joules per second. If we take the energy per bit as 0.693 kT, T as 3.5°K, and use the value of Boltzmann's constant given on page 188, we conclude that our communication system can send us over 800,000 bits per second, a very useful amount of information.

What about an optical communication system? Let us assume a wavelength of 6×10^{-7} meters, corresponding to a frequency of 5×10^{14} cycles per second. This is visible light. We will assume somewhat smaller antennas (lenses or mirrors), a transmitting antenna one meter square, with an area of 1 square meter and a receiving antenna 10 meters square, with an area of 100 square meters. Again, we will assume a transmitted power of 10 watts. The effective optical "temperature" of space, that is, the total starlight, is a little uncertain; we will take it as 350°K. We calculate a transmission capacity of 800 billion bits per second for our optical link.

If we receive 800 billion bits per second, we must receive 100 bits per photon. It seems unlikely that this is possible. But even if we received only one bit per photon we would receive 8 billion bits/second. Optical communication may be the best way to communicate over long distances in space.

From the point of view of information theory, the most interesting relation between physics and information theory lies in the evaluation of the unavoidable limitations imposed by the laws of physics on our ability to communicate. In a very fundamental sense, this is concerned with the limitations imposed by Johnson noise and quantum effects. It also, however, includes limitations imposed by atmospheric turbulence and by fluctuations in the ionosphere, which can distort a signal in a way quite different from adding noise to it. Many other examples of this sort of relation of physics to information theory could be unearthed.

Physicists have thought of a connection between physics and communication theory which has nothing to do with the fundamental problem that communication theory set out to solve, that is, the possibilities of the limitations of efficient encoding in transmitting information over a noisy channel. Physicists propose to use the idea of the transmission of information in order to show the impossibility of what is called *a perpetual-motion machine of the second kind.* As a matter of fact, this idea preceded the invention of communication theory in its present form, for L. Szilard put forward such ideas in 1929.

Some perpetual-motion machines purport to create energy; this violates the first law of thermodynamics, this is, the conservation of energy.

Other perpetual-motion machines purport to convert the disorganized energy of heat in matter or radiation which is all at the same temperature into ordered energy, such as the rotation of a flywheel. The rotating flywheel could, of course, be used to drive a refrigerator which would cool some objects and heat others. Thus, this sort of perpetual motion could, without the use of additional organized energy, transfer the energy of heat from cold material to hot material.

The second law of thermodynamics can be variously stated: that heat will not flow from a cold body to a hot body without the expenditure of organized energy or that the entropy of a system never decreases. The second sort of perpetual-motion machine violates the second law of thermodynamics.

One of the most famous perpetual-motion machines of this second kind was invented by James Clerk Maxwell. It makes use of a fictional character called *Maxwell's demon.*

I have pictured Maxwell's demon in Figure X-2. He inhabits a divided box and operates a small door connecting the two chambers of the box. When he sees a fast molecule heading toward the door from the far side, he opens the door and lets it into his side. When he sees a slow molecule heading toward the door from his side he lets it through. He keeps slow molecules from entering his side and fast molecules from leaving his side. Soon, the gas in his side is made up of fast molecules. It is hot, while the gas on the other side is made up of slow molecules and it is cool. Maxwell's demon makes heat flow from the cool chamber to the hot chamber. I have shown him operating the door with one hand and thumbing his nose at the second law of thermodynamics with his other hand.

Maxwell's demon has been a real puzzler to those physicists who have not merely shrugged him off. The best general objection we can raise to him is that, since the demon's environment is at thermal equilibrium, the only light present is the random electromagnetic radiation corresponding to thermal noise, and this is so chaotic that the demon can't use it to see what sort of molecules are coming toward the door.

We can think of other versions of Maxwell's demon. What about putting a spring door between the two chambers, for instance? A molecule hitting such a door from one side can open it and go through; one hitting it from the other side can't open it at all. Won't we end up with all the molecules and their energy on the side into which the spring door opens?

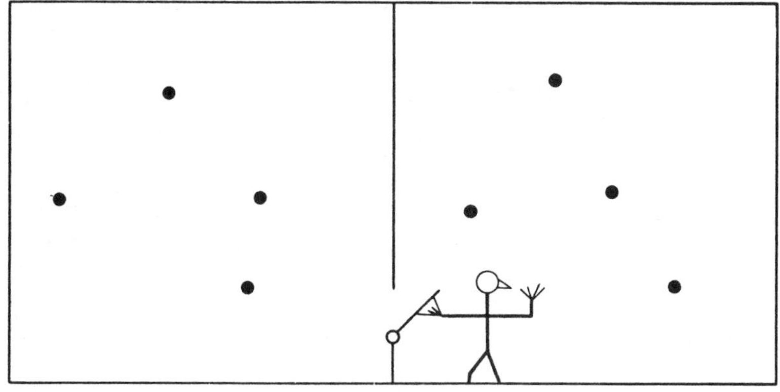

Fig. X-2

One objection which can be raised to the spring door is that, if the spring is strong, a molecule can't open the door, while, if the spring is weak, thermal energy will keep the door continually flapping, and it will be mostly open. Too, a molecule will give energy to the door in opening it. Physicists are pretty well agreed that such mechanical devices as spring doors or delicate ratchets can't be used to violate the second law of thermodynamics.

Arguing about what will and what won't work is a delicate business. An ingenious friend fooled me completely with his machine until I remembered that any enclosure at thermal equilibrium must contain random electromagnetic radiation as well as molecules. However, there is one simple machine which, although it is frictionless, ridiculous, and certainly inoperable in any practical sense, is, I believe, not physically impossible in the very special sense in which physicists use this expression. This machine is illustrated in Figure X-3.

The machine makes use of a cylinder *C* and a frictionless piston *P*. As the piston moves left or right, it raises one of the little pans *p* and lowers the other. The piston has a door in it which can be opened or closed. The cylinder contains just one molecule *M*. The whole device is at a temperature *T*. The molecule will continually gain and lose energy in its collisions with the walls, and it will have an average energy proportional to the temperature.

When the door in the piston is open, no work will be done if we move the piston slowly to the right or to the left. We start by centering the piston with the door open. We clamp the piston in

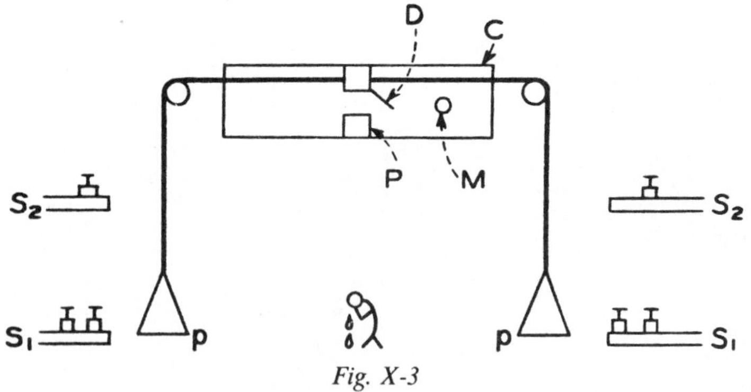

Fig. X-3

the center and close the door. We then observe which side of the piston the molecule is on. When we have found out which side of the piston the molecule is on, we put a little weight from low shelf S_1 onto the pan on the same side as the molecule and unclamp the piston. The repeated impact of the molecule on the piston will eventually raise the weight to the higher shelf S_2, and we take the weight off and put it on this higher shelf. We then open the door in the piston, center it, and repeat the process. Eventually, we will have lifted an enormous number of little weights from the lower shelves S_1 to the upper shelves S_2. We have done organized work by means of disorganized thermal energy!

How much work have we done? It is easily shown that the average force F which the molecule exerts on the piston is

$$F = \frac{\mathrm{k}T}{L} \tag{10.10}$$

Here L is the distance from the piston to the end of the cylinder on the side containing the molecule. When we allow the molecule to push against the piston and slowly drive it to the end of the cylinder, so that the distance is doubled, the *most* work W that the molecule can do is

$$W = 0.693\ \mathrm{k}T \tag{10.11}$$

Actually, in lifting a constant weight the work done will be less, but 10.11 represents the limit. Did we get this free?

Not quite! When we have centered the piston and closed the door it is equally likely that we will find the molecule in either half of the cylinder. In order to know which pan to put the weight on, we need one bit of information, specifying which side the molecule is on. To make the machine run we must receive this information in a system which is at a temperature T. What is the very least energy needed to transmit one bit of information at the temperature T? We have already computed this; from equation 10.6 we see that it is exactly 0.693 kT joule, just equal to the most energy the machine can generate. In principle, this applies to the quantum case, if we do the best that is possible. Thus, we use up all the output of the machine in transmitting enough information to make the machine run!

It's useless to argue about the actual, the attainable, as opposed to the limiting efficiency of such a machine; the important thing is that even at the very best we could only break even.

We have now seen in one simple case that the transmission of information in the sense of communication theory can enable us to convert thermal energy into mechanical energy. The *bit* which measures amount of information used is the unit in terms of which the entropy of a message source is measured in communication theory. The entropy of thermodynamics determines what part of existing thermal energy can be turned into mechanical work. It seems natural to try to relate the entropy of thermodynamics and statistical mechanics with the entropy of communication theory.

The entropy of communication theory is a measure of the uncertainty as to what message, among many possible messages, a message source will actually produce on a given occasion. If the source chooses a message from among m equally probable messages, the entropy in bits per message is the logarithm to the base 2 of m; in this case it is clear that such messages can be transmitted by means of $\log m$ binary digits per message. More generally, the *importance* of the entropy of communication theory is that it measures directly the average number of binary digits required to transmit messages produced by a message source.

The entropy of statistical mechanics is the uncertainty as to what state a physical *system* is in. It is assumed in statistical mechanics that all states of a given total energy *are* equally probable. The entropy of statistical mechanics is Boltzmann's constant times the logarithm to the base e of the number of possible states. This entropy has a wide importance in statistical mechanics. One matter of importance is that the free energy, which we will call *F.E.*, is given by

$$F.E. = E - HT \qquad (10.12)$$

Here E is the total energy, H is the entropy, and T is the temperature. The free energy is the part of the total energy which, ideally, can be turned into organized energy, such as the energy of a lifted weight.

In order to understand the entropy of statistical mechanics, we have to say what a physical system is, and we will do this by citing a few examples. A physical system can be a crystalline solid, a closed vessel containing water and water vapor, a container filled with gas, or any other substance or collection of substances. We will consider such a system when it is at equilibrium, that is, when it has settled down to a uniform temperature and when any physical or chemical changes that may tend to occur at this temperature have gone as far as they will go.

As a particular example of a physical system, we will consider and idealized gas made up of a lot of little, infinitely small particles, whizzing around every which way in a container.

The *state* of such a system is a complete description, or as complete a description as the laws of physics allow, of the positions and velocities of all of these particles. According to classical mechanics (Newton's laws of motion), each particle can have any velocity and energy, so there is an uncountably infinite number of states, as there is such an uncountable infinity of points in a line or a square. According to quantum mechanics, there is an infinite but countable number of states. Thus, the classical case is analogous to the difficult communication theory of continuous signals, while the more exact quantum case is analogous to the communication theory of discrete signals which are made up of a countable set of distinct, different symbols. We have dealt with the theory of discrete signals at length in this book.

According to quantum mechanics, a particle of an idealized gas can move with only certain energies. When it has one of these allowed energies, it is said to occupy a particular *energy level.* How large will the entropy of such a gas be? If we increase the volume of the gas, we increase the number of energy levels within a given energy range. This increases the number of states the system can be in at a given temperature, and hence it increases the entropy. Such an increase in entropy occurs if a partition confining a gas to a portion of a container is removed and the gas is allowed to expand suddenly into the whole container.

If the temperature of a gas of constant volume is increased, the particles can occupy energy levels of higher energy, so more combinations of energy levels can be occupied; this increases the number of states, and the entropy increases.

If a gas is allowed to expand against a slowly moving piston and no heat is added to the gas, the number of energy levels in a given energy range increases, but the temperature of the gas decreases just enough so as to keep the number of states and the entropy the same.

We see that for a given temperature, a gas confined to a small volume has less entropy than the same gas spread through a larger volume. In the case of the one-molecule gas of Figure X-3, the entropy is less when the door is closed and the molecule is confined to the space on one side of the piston. At least, the entropy is less if we *know* which side of the piston the molecule is on.

We can easily compute the decrease in entropy caused by halving the volume of an ideal, one-molecule, classical gas at a given temperature. In halving the volume we halve the number of states, and the entropy changes by an amount

$$\mathrm{k} \log_e \tfrac{1}{2} = -0.693\ \mathrm{k}$$

The corresponding change in free energy is the negative of T times this change in entropy, that is,

$$0.693\ \mathrm{k}T$$

This is just the work that, according to 10.11, we can obtain by halving the volume of the one-molecule gas and then letting it expand against the piston until the volume is doubled again. Thus, computing the change in free energy is one way of obtaining 10.11.

In reviewing our experience with the one-molecule heat engine in this light, we see that we must transmit one bit of information in order to specify on which side of the piston the molecule is. We must transmit this information against a background of noise corresponding to the uniform temperature T. To do this takes 0.693 kT joule of energy.

Because we now know that the molecule is definitely on a particular side of the piston, the entropy is 0.693 k less than it would be if we were uncertain as to which side of the piston the molecule was on. This reduction of entropy corresponds to an increase in free energy of 0.693 kT joule. This free energy we can turn into work by allowing the piston to move slowly to the unoccupied end of the cylinder while the molecule pushes against it in repeated impacts. At this point the entropy has risen to its original value, and we have obtained from the system an amount of work which, alas, is just equal to the minimum possible energy required to transmit the information which told us on which side of the piston the molecule was.

Let us now consider a more complicated case. Suppose that a physical system has at a particular temperature a total of m states. Suppose that we divide these states into n equal groups. The number of states in each of these groups will be m/n.

Suppose that we regard the specification as to which one of the n groups of states contains the state that the system is in as a message source. As there are n equally likely groups of states, the communication-theory entropy of the source is log n bits. This means that it will take n binary digits to specify the particular group of states which contains the state the system is actually in. To transmit this information at a temperature T requires at least

$$.693\ \mathrm{k}T \log n = \mathrm{k}T \log_e n$$

joule of energy. That is, the energy required to transmit the message is proportional to the communication-theory entropy of the message source.

If we know merely that the system is in one of the total of m states, the entropy is

$$\mathrm{k} \log_e m$$

If we are sure that the system is in one particular group of states containing only m/n states (as we are after transmission of the information as to which state the system is in), the entropy is

$$\mathrm{k} \log_e \frac{m}{n} = \mathrm{k}\ (\log_e m - \log_e n)$$

The change in entropy brought about by information concerning which one of the n groups of states the system is in is thus

$$-\mathrm{k} \log_e n$$

The corresponding increase in free energy is

$$\mathrm{k}T \log_e n$$

But this is just equal to the least energy necessary to transmit the information as to which group of states contains the state the system is in, the information that led to the decrease in entropy and the increase in free energy.

We can regard any process which specifies something concerning which state a system is in as a message source. This source generates a message which reduces our uncertainty as to what state the system is in. Such a source has a certain communication-theory entropy per message. This entropy is equal to the number of binary digits necessary to transmit a message generated by the source. It takes a particular energy per binary digit to transmit the message against a noise corresponding to the temperature T of the system.

The message reduces our uncertainty as to what state the system is in, thus reducing the entropy (of statistical mechanics) of the system. The reduction of entropy increases the free energy of the system. But, the increase in free energy is just equal to the minimum energy necessary to transmit the message which led to the increase of free energy, an energy proportional to the entropy of communication theory.

This, I believe, is the relation between the entropy of communication theory and that of statistical mechanics. One pays a price for information which leads to a reduction of the statistical-mechanical entropy of a system. This price is proportional to the communication-theory entropy of the message source which produces the information. It is always just high enough so that a perpetual motion machine of the second kind is impossible.

We should note, however, that a message source which generates messages concerning the state of a physical system is one very particular and peculiar kind of message source. Sources of English text or of speech sounds are much more common. It seems irrelevant to relate such entropies to the entropy of physics, except perhaps through the energy required to transmit a bit of information under highly idealized conditions.

There is one very odd implication of what we have just covered. Clearly, the energy needed to transmit information about the state of a physical system keeps us from ever knowing the past in complete detail. If we can never know the past fully, can we declare that the past is indeed unique? Or, is this a sensible question?

To summarize, in this chapter we have considered some of the problems of communicating electrically in our actual physical world. We have seen that various physical phenomena, including lightning and automobile ignition systems, produce electrical disturbances or noise which are mixed with the electrical signals we use for the transmission of messages. Such noise is a source of error in the transmission of signals, and it limits the rate at which we can transmit information when we use a particular signal power and band width.

The noise emitted by hot bodies (and any body is hot to a degree if its temperature is greater than absolute zero) is a particularly simple, universal, unavoidable noise which is important in all communication. But, at extremely high frequencies there are quantum effects as well as Johnson or thermal noise. We have seen how these affect communication in the limit of infinite bandwidth, but have no quantum analog of expression 10.4.

The use of the term entropy in both physics and communication theory has raised the question of the relation of the two entropies. It can be shown in a simple case that the limitation imposed by thermal noise on the transmission of information results in the failure of a machine designed to convert the chaotic energy of heat into the organized energy of a lifted weight. Such a machine, if it succeeded, would violate the second law of thermodynamics. More generally, suppose we regard a source of information as to what state a system is in as a message source. The information-theory entropy of this source is a measure of the energy needed to transmit a message from the source in the presence of the thermal noise which is necessarily present in the system. The energy used in transmitting such a message is as great as the increase in free energy due to the reduction in physical entropy which the message brings about.

CHAPTER XI *Cybernetics*

SOME WORDS HAVE a heady quality; they conjure up strong feelings of awe, mystery, or romance. *Exotic* used to be Dorothy Lamour in a sarong. Just what it connotes currently I don't know, but I am sure that its meaning, foreign, is pale by comparison. *Palimpsest* makes me think of lost volumes of Solomon's secrets or of other invaluable arcane lore, though I know that the word means nothing more than a manuscript erased to make room for later writing.

Sometimes the spell of a word or expression is untainted by any clear and stable meaning, and through all the period of its currency its magic remains secure from commonplace interpretations. *Tao, élan vital,* and *id* are, I think, examples of this. I don't believe that *cybernetics* is quite such a word, but it does have an elusive quality as well as a romantic aura.

The subtitle of Wiener's book, *Cybernetics,* is *Control and Communication in the Animal and the Machine.* Wiener derived the word from the Greek for steersman. Since the publication of Wiener's book in 1948, *cybernetics* has gained a wide currency. Further, if there is cybernetics, then someone must practice it, and *cyberneticist* has been anonymously coined to designate such a person.

What *is* cybernetics? If we are to judge from Wiener's book it includes at least information theory, with which we are now reasonably familiar; something that might be called smoothing, filtering, detection and prediction theory, which deals with finding the presence of and predicting the future value of signals, usually in the presence of noise; and negative feedback and servomechanism theory, which Wiener traces back to an early treatise on the governor (the device that keeps the speed of a steam engine constant) published by James Clerk Maxwell in 1868. We must, I think, also include another field which may be described as automata and complicated machines. This includes the design and programming of digital computers.

Finally, we must include any phenomena of life which resemble anything in this list or which embody similar processes. This brings to mind at once certain behavioral and regulatory functions of the body, but Wiener goes much further. In his second autobiographical volume, *I Am a Mathematician,* he says that sociology and anthropology are primarily sciences of communication and therefore fall under the general head of cybernetics, and he includes, as a special branch of sociology, economics as well.

One could doubt Wiener's sincerity in all this only with difficulty. He had a grand view of the importance of a statistical approach to the whole world of life and thought. For him, a current which stems directly from the work of Maxwell, Boltzmann, and Gibbs swept through his own to form a broad philosophical sea in which we find even the ethics of Kierkegaard.

The trouble is that each of the many fields that Wiener drew into cybernetics has a considerable scope in itself. It would take many thousands of words to explain the history, content, and prospects of any one of them. Lumped together, they constitute not so much an exciting country as a diverse universe of overwhelming magnitude and importance.

Thus, few men of science regard themselves as cyberneticists. Should you set out to ask, one after another, each person listed in *American Men of Science* what his field is, I think that few would reply cybernetics. If you persisted and asked, "Do you work in the field of cybernetics?" a man concerned with communication, or with complicated automatic machines such as computers, or with some parts of experimental psychology or neurophysiology would look at you and speculate on your background and intentions. If he decided that you were a sincere and innocent outsider, who would in any event never get more than a vague idea of his work, he might well reply, "yes."

So far, in this country the word *cybernetics* has been used most extensively in the press and in popular and semiliterary, if not semiliterate, magazines. I cannot compete with these in discussing the grander aspects of cybernetics. Perhaps Wiener has done that best himself in *I Am a Mathematician*. Even the more narrowly technical content of the fields ordinarily associated with the word cybernetics is so extensive that I certainly would never try to explain it all in one book, even a much larger book than this.

In this one chapter, however, I propose to try to give some small idea of the nature of the different technical matters which come to mind when cybernetics is mentioned. Such a brief résumé may perhaps help the reader in finding out whether or not he is interested in cybernetics and indicate to him what sort of information he should seek in order to learn more about it.

Let us start with the part of cybernetics that I have called smoothing, filtering, and prediction theory, which is an extremely important field in its own right. This is a highly mathematical subject, but I think that some important aspects of it can be made pretty clear by means of a practical example.

Suppose that we are faced with the problem of using radar data to point a gun so as to shoot down an airplane. The radar gives us a sequence of measurements of position each of which is a little in error. From these measurements we must deduce the course and the velocity of the airplane, so that we can predict its position at some time in the future, and by shooting a shell to that position, shoot the plane down.

Suppose that the plane has a constant velocity and altitude. Then the radar data on its successive locations might be the crosses of Figure XI-1. We can by eye draw a line *AB*, which we would guess to represent the course of the plane pretty well. But how are we to tell a machine to do this?

If we tell a computing machine, or "computer," to use just the last and next-to-last pieces of radar data, represented by the points *L* and *NL*, it can only draw a line through these points, the dashed line $A'B'$. This is clearly in error. In some way, the computer must use earlier data as well.

The simplest way for the computer to use the data would be to give an equal weight to all points. If it did this and fitted a straight

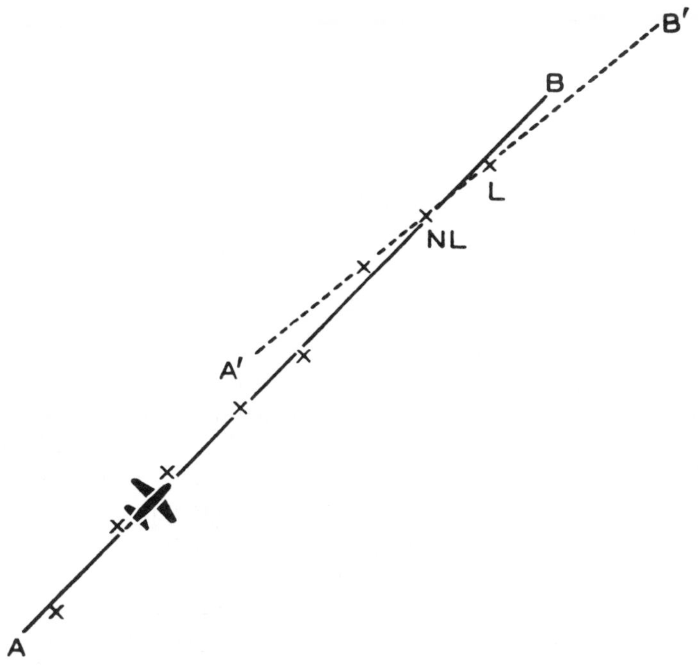

Fig. XI-1

line to all the data taken together, it might get a result such as that shown in Figure XI-2. Clearly, the airplane turned at point *T,* and the straight line *AB* that the computer computed has little to do with the path of the plane.

We can seek to remedy this by giving more importance to recent data than to older data. The simplest way to do this is by means of *linear* prediction. In making a linear prediction, the computer takes each piece of radar data (a number representing the distance north or the distance east from the radar, for instance) and multiplies it by another number. This other number depends on how recent the piece of data is; it will be a smaller number for an old piece of data than for a recent one. The computer then adds up all the products it has obtained and so produces a predicted piece of data (for instance, the distance north or east of the radar at some future time).

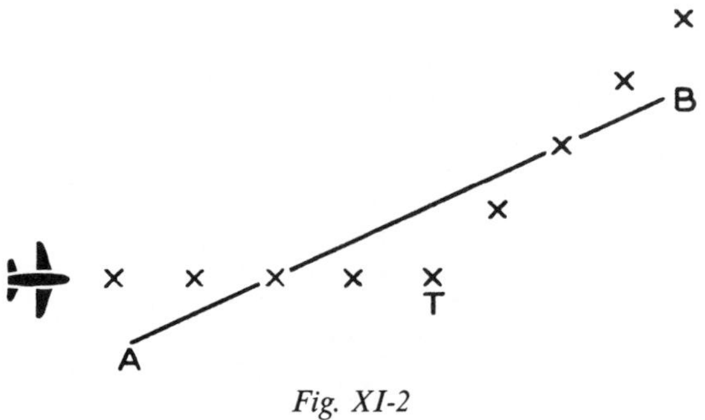

Fig. XI-2

The result of such prediction might be as shown in Figure XI-3. Here a linear method has been used to estimate a new position and direction each time a new piece of radar data, represented by a cross, becomes available. Until another piece of data becomes available, the predicted path is taken as a straight line proceeding from the estimated location in the estimated direction. We see that it takes a long time for the computer to take into account the fact that the plane has turned at the point *T*, despite the fact that we are sure of this by the time we have looked at the point next after *T*.

A linear prediction can make good use of old data, but, if it does this, it will be slow to respond to new data which is inconsistent with the old data, as the data obtained after an airplane turns will be. Or a linear prediction can be quick to take new data strongly into account, but in this case it will not use old data effectively, even when the old data is consistent with the new data.

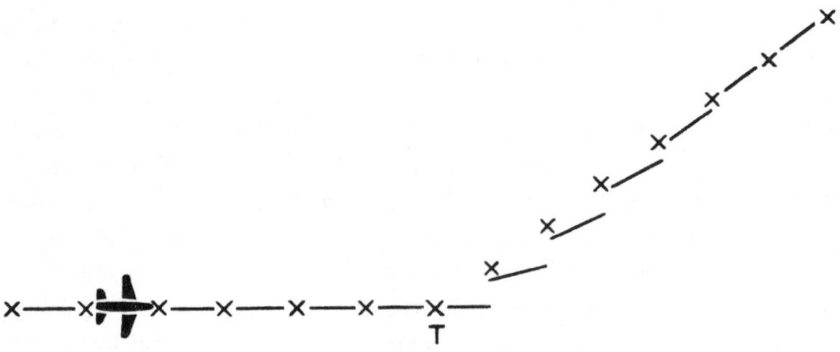

T

Fig. XI-3

To predict well even when circumstances change (as when the airplane turns) we must use *nonlinear* prediction. Nonlinear prediction includes all methods of prediction in which we don't merely multiply each piece of data used by a number depending on how old the data is and then add the products.

As a very simple example of nonlinear prediction, suppose that we have two different linear predictors, one of which takes into account the last 100 pieces of data received, and the other of which takes into account only the last ten pieces of data received. Suppose that we use each predictor to estimate the next piece of data which will be received. Suppose that we compare this next piece of data with the output of each predictor. Suppose that we make use of predictions based on 100 past pieces of data only when, three times in a row, such predictions agree with each new piece of data better than predictions based on ten past pieces of data. Otherwise, we assume that the aircraft is maneuvering in such a way as to make long-past data useless, and we use predictions based on ten past pieces of data. This way of arriving at a final prediction is nonlinear because the prediction is not arrived at simply by multiplying each past piece of data by a number which depends only on how old the data is. Instead, the use we make of past data depends on the nature of the data received.

More generally, there are endless varieties of nonlinear prediction. In fact, nonlinear prediction, and other nonlinear processes as well, are the overwhelming total of all very diverse means after the simplest category, linear prediction and other linear processes, have been excluded. A great deal is known about linear prediction, but very little is known about nonlinear prediction.

This very special example of predicting the position of an airplane has been used merely to give a concrete sense of something which might well seem almost meaningless if it were stated in more abstract terms. We might, however, restate the broader problem, which has been introduced in a more general way.

Let us imagine a number of possible signals. These signals might consist of things as diverse as the possible paths of airplanes or the possible different words that a man may utter. Let us also imagine some sort of noise or distortion. Perhaps the radar data is inexact, or perhaps the man speaks in a noisy room. We are required to estimate some aspect of the correct signal: the present or future position of the airplane, the word the man just spoke, or the word that he will speak next. In making this judgment we have some statistical knowledge of the signal. This might concern what airplane paths are most likely, or how often turns are made, or how sharp they are. It might include what words are most common and how the likelihood of their occurrence depends on preceding words. Let us suppose that we also have similar statistics concerning noise and distortion.

We see that we are considering exactly the sort of data that are used in communication theory. However, given a source of data and a noisy channel, the communication theorist asks how he can best encode messages from the source for transmission over the channel. In prediction, given a set of signals distorted by noise, we ask, how do we best detect the true signal or estimate or predict some aspect of it, such as its value at some future time?

The armory of prediction consists of a general theory of linear prediction, worked out by Kolmogoroff and Wiener, and mathematical analyses of a number of special nonlinear predictors. I don't feel that I can proceed very profitably beyond this statement, but I can't resist giving an example of a theoretical result (due to David Slepian, a mathematician) which I find rather startling.

Let us consider the case of a faint signal which may or may not be present in a strong noise. We want to determine whether or not the signal is present. The noise and the signal might be voltages or sound pressures. We assume that the noise and the signal have been combined simply by adding them together. Suppose further that the signal and the noise are ergodic (*see* Chapter III) and that they are band limited—that is, they contain no frequencies outside of a specified frequency range. Suppose further that we know exactly the frequency spectrum of the noise, that is, what fraction of the noise power falls in every small range of frequencies. Suppose that the frequency spectrum of the signal is different from this. Slepian has shown that if we could measure the over-all voltage (or sound pressure) of the signal plus noise *exactly* for every instant in any interval of time, however short the interval is, we could infallibly tell whether or not the signal was present along with the noise, no matter how faint the signal might be. This is a sound theoretical, not a useful practical, conclusion. However, it has been a terrible shock to a lot of people who had stated quite positively that, if the signal was weak enough (and they stated just how weak), it could not be detected by examining the signal plus noise for any particular finite interval of time.

Before leaving this general subject, I should explain why I described it in terms of *filtering* and *smoothing* as well as *prediction* and *detection.* If the noise mixed with a signal has a frequency spectrum different from that of the signal, we will help to separate the signal from the noise by using an electrical filter which cuts down on the frequencies which are strongly present in the noise with respect to the frequencies which are strongly present in the signal. If we use a filter which removes most or all high frequency components (which vary rapidly with time), the output will not vary so abruptly with time as the input; we will have *smoothed* the combination of signal and noise.

So far, we have been talking about operations which we perform on a set of data in order to estimate a present or future signal or to detect a signal. This is, of course, for the purpose of doing something.

We might, for instance, be flying an airplane in pursuit of an enemy plane. We might use a radar to see the enemy plane. Every time we take an observation, we might move the controls of the plane so as to head toward the enemy.

A device which acts continually on the basis of information to attain a specified goal in the face of changes is called a *servo-mechanism*. Here we have an important new element, for the radar data measures the position of the enemy plane with respect to our plane, and the radar data is used in determining how the position of our plane is to be changed. The radar data is *fed back* in such a way as to alter the nature of radar data which will be obtained later (because the data are used to alter the position of the plane from which new radar data are taken). The feedback is called *negative feedback*, because it is so used as to decrease rather than to increase any departure from a desired behavior.

We can easily think of other examples of negative feedback. The governor of a steam engine measures the speed of the engine. This measured value is used in opening or closing the throttle so as to keep the speed at a predetermined value. Thus, the result of the measurement of speed is fed back so as to change the speed. The thermostat on the wall measures the temperature of the room and turns the furnace off or on so as to maintain the temperature at a constant value. When we walk carrying a tray of water, we may be tempted to watch the water in the tray and try to tilt the tray so as to keep the water from spilling. This is often disastrous. The more we tilt the tray to avoid spilling the water, the more wildly the water may slosh about. When we apply feedback so as to change a process on the basis of its observed state, the over-all situation may be *unstable*. That is, instead of reducing small deviations from the desired goal, the control we exert may make them larger.

This is a particularly hazardous matter in feedback circuits. The thing we do to make corrections most complete and perfect is to make the feedback stronger. But this is the very thing that tends to make the system unstable. Of course, an unstable system is no good. An unstable system can result in such behavior as an airplane or missile veering wildly instead of following the target, the temperature of a room rising and falling rapidly, an engine racing or coming to a stop, or an amplifier producing a singing output of high amplitude when there is no input.

The stability of negative-feedback systems has been studied extensively, and a great deal is known about linear negative-feedback systems, in which the present amplitude is the sum of past amplitudes multiplied by numbers depending only on remoteness from the present.

Linear negative-feedback systems are either stable or unstable, regardless of the input signal applied. Nonlinear feedback systems can be stable for some inputs but unstable for others. A shimmying car is an example of a nonlinear system. It can be perfectly stable at a given speed on a smooth road, and yet a single bump can start a shimmy which will persist indefinitely after the bump has been passed.

Oddly enough, most of the early theoretical work on negative-feedback systems was done in connection with a device which has not yet been described. This is the *negative feedback amplifier*, which was invented by Harold Black in 1927.

The *gain* of an amplifier is the ratio of the output voltage to the input voltage. In telephony and other electronic arts, it is important to have amplifiers which have a very nearly constant gain. However, vacuum tubes and transistors are imperfect devices. Their gain changes with time, and the gain can depend on the strength of the signal. The negative feedback amplifier reduces the effect of such changes in the gain of vacuum tubes or transistors.

We can see very easily why this is so by examining Figure XI-4. At the top we have an ordinary amplifier with a gain of ten times. If we put in 1 volt, as shown by the number to the left, we get out 10 volts, as shown by the number to the right. Suppose the gain of the amplifier is halved, so that the gain is only five times, as shown next to the top. The output also falls to one half, or 5 volts, in just the same ratio as the gain fell.

The third drawing from the top shows a negative feedback amplifier designed to give a gain of ten times. The upper box has

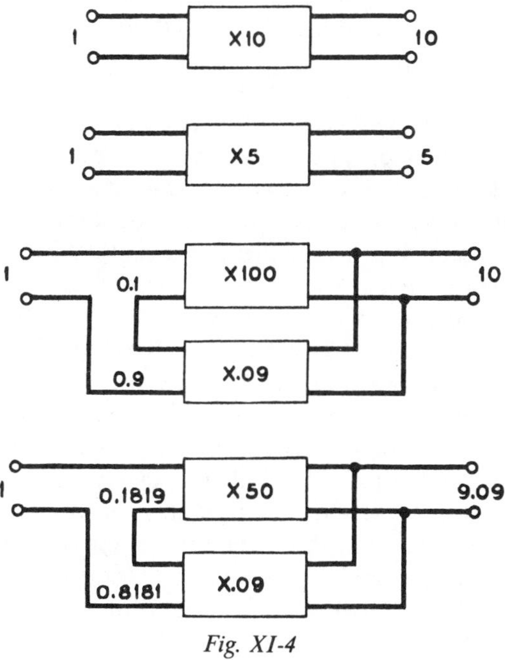

Fig. XI-4

a high gain of one hundred times. The output of this box is connected to a very accurate voltage-dividing box, which contains no tubes or transistors and does not change with time or signal level. The input to the upper box consists of the input voltage of 1 volt less the output of the lower box, which is 0.09 times the output voltage of 10 volts; this is, of course, 0.9 volt.

Now, suppose the tubes or transistors in the upper box change so that they give a gain of only fifty times instead of one hundred times; this is shown at the bottom of Figure XI-4. The numbers given in the figure are only approximate, but we see that when the gain of the upper box is cut in half the output voltage falls only about 10 per cent. If we had used a higher gain in the upper box the effect would have been even less.

The importance of negative feedback can scarcely be overestimated. Negative feedback amplifiers are essential in telephonic communication. The thermostat in your home is an example of negative feedback. Negative feedback is used to control chemical processing plants and to guide missiles toward airplanes. The automatic pilot of an aircraft uses negative feedback in keeping the plane on course.

In a somewhat broader sense, I use negative feedback from eye to hand in guiding my pen across the paper, and negative feedback from ear to tongue and lips in learning to speak or in imitating the voice of another. The animal organism makes use of negative feedback in many other ways. This is how it maintains its temperature despite changes in outside temperature, and how it maintains constant chemical properties of the blood and tissues. The ability of the body to maintain a narrow range of conditions despite environmental changes has been called *homeostasis.*

G. Ross Ashby, one of the few self-acknowledged cyberneticists, built a machine called a *homeostat* to demonstrate features of adjustment to environment which he believes to be characteristic of life. The homeostat is provided with a variety of feedback circuits and with two means for changing them. One is under the control of the homeostat; the other is under the control of a person who acts as the machine's "environment." If the machine's circuits are so altered by changes of its "environment" as to make it unstable, it readjusts its circuits by trial and error so as to attain stability again.

We may if we wish liken this behavior of the homeostat to that of a child who first learns to walk upright without falling and then learns to ride a bicycle without falling or to many other adjustments we make in life. In his book *Cybernetics,* Wiener puts great emphasis on negative feedback as an element of nervous control and on its failure as an explanation of disabilities, such as tremors of the hand, which are ascribed to failures of a negative feedback system of the body.

We have so far discussed three constituents of cybernetics: information theory, detection and prediction, including smoothing and filtering, and negative feedback, including servomechanisms and negative feedback amplifiers. We usually also associate electronic computers and similar complex devices with cybernetics. The word *automata* is sometimes used to refer to such complicated machines.

One can find many precursors of today's complicated machines in the computers, automata, and other mechanisms of earlier centuries, but one would add little to his understanding of today's complex devices by studying these precursors. Human beings learn by doing and by thinking about what they have done. The opportunities for doing in the field of complicated machines have been enhanced immeasurably beyond those of previous centuries, and the stimulus to thought has been wonderful to behold.

Recent advances in complicated machines might well be traced to the invention of automatic telephone switching late in the last century. Early telephone switching systems were of a primitive, *step-by-step* form, in which a mechanism set up a new section of a link in a talking path as each digit was dialed. From this, switching systems have advanced to become *common-control* systems. In a common-control switching system, the dialed number does not operate switches directly. It is first stored, or represented electrically or mechanically, in a particular portion of the switching system. Electrical apparatus in another portion of the switching system then examines different electrical circuits that could be used to connect the calling party to the number called, until it finds one that is not in use. This free circuit is then used to connect the calling party to the called party.

Modern telephone switching systems are of bewildering complexity and overwhelming size. Linked together to form a nationwide telephone network which allows dialing calls clear across the country, they are by far the most complicated construction of man. It would take many words to explain how they perform even a few of their functions. Today, a few pulls of a telephone dial will cause telephone equipment to seek out the most economical available path to a distant telephone, detouring from city to city if direct paths are not available. The equipment will establish a connection, ring the party, time the call, and record the charge in suitable units, and it will disconnect the circuits when a party hangs up. It will also report malfunctioning of its parts to a central location, and it continues to operate despite the failure of a number of devices.

One important component of telephone switching systems is the electric *relay*. The principal elements of a relay are an electromagnet, a magnetic bar to which various movable contacts are attached, and fixed contacts which the movable contacts can touch, thus closing circuits. When an electric current is passed through the coil of the electromagnet of the relay, the magnetic bar is attracted and moves. Some moving contacts move away from the corresponding fixed contacts, opening circuits; other moving contacts are brought into contact with the corresponding fixed contacts, closing circuits.

In the thirties, G. R. Stibitz of the Bell Laboratories applied the relays and other components of the telephone art to build a *complex calculator,* which could add, subtract, multiply, and divide complex numbers. During World War II, a number of more complicated relay computers were built for military purposes by the Bell Laboratories, while, in 1941, Howard Aiken and his coworkers built their first relay computer at Harvard.

An essential step in increasing the speed of computers was taken shortly after the war when J. P. Eckert and J. W. Mauchly built the Eniac, a vacuum tube computer, and more recently transistors have been used in place of vacuum tubes.

Thus, it was an essential part of progress in the field of complex machines that it became possible to build them and that they were built, first by using relays and then by using vacuum tubes and transistors.

The building of such complex devices, of course, involved more than the existence of the elements themselves; it involved their interconnection to do particular functions such as multiplication and division. Stibitz's and Shannon's application of Boolean algebra, a branch of mathematical logic, to the description and design of logic circuits has been exceedingly important in this connection.

Thus, the existence of suitable components and the art of interconnecting them to carry out particular functions provided, so to speak, the body of the complicated machine. The organization, the spirit, of the machine is equally essential, though it would scarcely have evolved in the absence of the body.

Stibitz's complex calculator was almost spiritless. The operator sent it pairs of complex numbers by teletype, and it cogitated and sent back the sum, difference, product, or quotient. By 1943, however, he had made a relay computer which received its instructions in sequence by means of a long paper tape, or *program,* which prescribed the numbers to be used and the sequences of operations to be performed.

A step forward was taken when it was made possible for the machine to refer back to an earlier part of the program tape on completing a part of its over-all task or to use subsidiary tapes to help it in its computations. In this case the computer had to make a *decision* that it had reached a certain point and then act on the basis of the decision. Suppose, for instance, that the computer was computing the value of the following series by adding up term after term:

$$1 - 1/3 + 1/5 - 1/7 + 1/9 - 1/11 + \dots$$

We might program the computer so that it would continue adding terms until it encountered a term which was less than 1/1,000,000 and then print out the result and go on to some other calculation. The computer could decide what to do next by subtracting the latest term computed from 1/1,000,000. If the answer was negative, it would compute another term and add it to the rest; if the answer was positive, it could print out the sum arrived at and refer to the program for further instructions.

The next big step in the history of computers is usually attributed to John von Neumann, who made extensive use of early computers in carrying out calculations concerning atomic bombs. Even early computers had *memories,* or *stores,* in which the numbers used in intermediate steps of a computation were retained for further processing and in which answers were stored prior to printing them out. Von Neumann's idea was to put the instructions, or program, of the machine, not on a separate paper tape, but right into the machine's *memory*. This made the instructions easily and flexibly available to the machine and made it possible for the machine to modify parts of its instructions in accordance with the results of its computations.

In old-fashioned desk calculating machines, decimal digits were stored on wheels which could assume any of ten distinct positions of rotation. In today's desk and hand calculators, numbers are stored in binary form in some of the tens of thousands of solid-state circuit elements of a large-scale integrated circuit chip. To retain numbers in such storage, electric power (a minute amount) must be supplied continually to the integrated circuit. Such storage of data is sometimes called volatile (the data disappears when power is interrupted). But, volatile storage can be made very reliable by guarding against interruption of power.

Magnetic core memory-arrays of tiny rings of magnetic material with wires threaded through them provide nonvolatile memory in computers larger than hand calculators and desk calculators. An interruption of power does not change the magnetization of the cores, but an electric transient due to faulty operation can change what is in memory.

Integrated circuit memory and core memory are *random access* memory. Any group of binary digits (a *byte* is 8 consecutive digits), or a *word* (of 8, 16, 32 or more digits) can be retrieved in a fraction of a microsecond by sending into the memory a sequence of binary digits constituting an *address*. Such memory is called *random access* storage.

In the early days, sequences of binary digits were stored as holes punched in paper tape. Now they are stored as sequences of magnetized patterns on magnetic tape or magnetic disks. Cassettes similar to or identical with audio cassettes provide cheap storage in small "recreational" computers. Storage on tape, disk or cassette is *sequential*; one must run past a lot of bits or words to get to the one you want. Sequential storage is slow compared with random access storage. It is used to store large amounts of data, or to store programs or data that will be loaded into fast, random access storage repeatedly. Tape storage is also used as *back up storage*, to preserve material stored in random access storage and disk storage that might be lost through computer malfunction.

Besides memory or storage, computers have arithmetic units which perform various arithmetical and logical operations, index registers to count operations, input and output devices (keyboards, displays and printers) control units which give instructions to other units and receive data from them, and can have many specialized pieces of hardware to do specialized chores such as doing floating point arithmetic, performing Fourier analyses or inverting matrices.

The user of a computer who wishes it to perform some useful or amusing task must write a program that tells the computer in explicit detail every elementary operation it must go through in attaining the desired result. Because a computer operates internally in terms of binary numbers, early programmers had to write out long sequences of binary instructions, laboriously and fallibly.

But, a computer can be used to translate sequences of letters and decimal digits into strings of binary digits, according to explicit rules. Moreover, subroutines can be written that will, in effect, cause the computer to carry out desirable operations (such as dividing, taking square root, and so on) that are not among the basic commands or functions provided by the hardware. Thus, assemblers or assembly language was developed. Assembly language is called machine language, though it is one step removed from the binary numbers that are stored in and direct the operations of the hardware.

When one writes a program in assembly language, he still has to write every step in sequence, though he can specify that various paths be taken through the program depending on the outcome of one step (whether a computed number is larger than, smaller than or equal to another, for example). But, the numbers he writes are in decimal form, and the instructions are easily remembered sequences of letters such as CLA (clear add, a command to set the accumulator to zero and add to it the number in a specified memory location).

Instructing a computer in machine (assembly) language is a laborious task. Computers are practical because they have operating systems, by means of which a few simple instructions will cause data to be read in, to be read out, perform other useful functions, and because programs (even operating systems themselves) are written in a higher-order language.

There are many higher-order languages. *Fortran* (formula translation) is one of the oldest and most persistent. It is adapted to numerical problems. A line of *Fortran* might read

40 IF SIN(X) < M THEN 80

The initial number 40 is the number of the program step. It instructs the computer to go to program step 80 if sin x is less than M.

Basic is a widely used language similar to but somewhat simpler than *Fortran*. The C language is useful in writing operating systems as well as in numerical calculations. Languages seem to be easiest to use and most efficient (in computer running time) when they are adapted to particular uses, such as numerical calculations or text manipulation or running simulations of electrical or mechanical or economic systems.

One gets from a program in a higher-order language to one in machine language instructions by means of a compiler, which translates a whole program, or by an interpreter, which translates the program line by line. Compilers are more efficient and more common. One writes programs by means of a text editor (or editing program) which allows one to correct mistakes—to revise the program without keying every symbol in afresh.

Today many children learn to program computers in grade school, more in high school and more still in college. You can't get through Dartmouth and some other colleges without learning to use a computer, even if you are a humanities major.

More and more, computers are used for tasks other than scientific calculations or keeping books. Such tasks include the control of spacecraft and automobile engines, the operation of chemical plants and factories, keeping track of inventory, making airline and hotel reservations, making word counts and concordances, analyzing X-ray data to produce a three-dimensional image (CAT = computer-aided tomography), composing (bad) music, producing attractive musical sounds, producing engaging animated movies, playing games, new and old, including chess, reading printed or keyed-in text aloud, and a host of other astonishing tasks. Through large-scale integration, simple computers have become so cheap that they are incorporated into attractive toys.

While children learn a useful amount of programming with little difficulty, great talent and great skill are required in order to get the most out of a computer or a computerlike device, or to accomplish a given task with a smaller and cheaper computer.

Today, far more money and human effort are spent in producing software (that is, programs of all sorts) than are spent on the hardware the programs run on. One talented programmer can accomplish what ten or a hundred merely competent programmers cannot do. But, there aren't enough talented programmers to go around. Indeed, programmers of medium competence are in short supply.

Even the most talented programmers haven't been able to make computers do some things. In principle, a computer can be programmed to do anything which the programmer understands in detail. Sometimes, of course, a computation may be too costly or too time-consuming to undertake. But often we don't really know in detail how to do something. Thus, things that a computer has not done so far include typing out connected speech, translating satisfactorily from one language to another, identifying an important and interesting mathematical theorem or composing interesting music.

Efforts to use the computer to perform difficult tasks that are not well understood has greatly stimulated human thought concerning such problems as the recognition of human words, the structure of language, the strategy of winning games, the structure of music and the processes of mathematical proof.

Further, the programming of computers to solve complicated and unusual problems has given us a new and objective criterion of understanding. Today, if a man says that he understands how a human being behaves in a given situation or how to solve a certain mathematical or logical problem, it is fair to insist that he demonstrate his understanding by programming a computer to imitate the behavior or to accomplish the task in question. If he is unable to do this, his understanding is certainly incomplete, and it may be completely illusory.

Will computers be able to think? This is a meaningless question unless we say what we mean by *to think.* Marvin Minsky, a free-wheeling mathematician who is much interested in computers and complex machines, proposed the following fable. A man beats everyone else at chess. People say, "How clever, how intelligent, what a marvelous mind he has, what a superb thinker he is." The man is asked, "How do you play so that you beat everyone?" He says, "I have a set of rules which I use in arriving at my next move." People are indignant and say, "Why that isn't thinking at all; it's just mechanical."

Minsky's conclusion is that people tend to regard as thinking only such things as they don't understand. I will go even further and say that people frequently regard as thinking almost any grammatical jumbling together of "important" words. At times I'd settle for a useful, problem-solving type of "thinking," even if it was mechanical. In any event, it seems likely that philosophers and humanists will manage to keep the definition of thinking perpetually applicable to human beings and a step ahead of anything a machine ever manages to do. If this makes them happy, it doesn't offend me at all. I do think, however, that it is probably impossible to specify a meaningful and explicitly defined goal which a man can attain and a computer cannot, even including the "imitation game" proposed by A. M. Turing, a British logician, in 1936.

In this game a man is in communication, say by teletype, with either a computer or a man, he doesn't know which. The man tries by means of questions to discover whether he is in touch with a man or a machine; the computer is programmed to deceive the man. Certainly, however, a computer programmed to play the imitation game with any chance of success is far beyond today's computers and today's art of programming, and it belongs to a very distant future, if to any.

We have seen that cybernetics is a very broad field indeed. It includes communication theory, to which we are devoting a whole book. It includes the complicated field of smoothing and prediction, which is so important in radar and in many other military applications. When we try to estimate the true position or the future position of an airplane on the basis of imperfect radar data, we are, according to Wiener, dealing with cybernetics. Even in using an electrical filter to separate noise of one frequency from signals of another frequency, we are invoking cybernetics.

It is in this general field that the contribution of Wiener himself was greatest, and his great work was a general theory of prediction by means of linear devices, which makes a prediction merely by multiplying each piece of data by a number which is smaller the older the data is and adding the products.

Another part of cybernetics is negative feedback. A thermostat makes use of negative feedback when it measures the temperature of a room and starts or stops the furnace in order to make the temperature conform to a specified value. The autopilots of airplanes use negative feedback in manipulating the controls in order to keep the compass and altimeter readings at assigned values. Human beings use negative feedback in controlling the motions of their hands to achieve certain ends.

Negative feedback devices can be unstable; the effect of the output can sometimes be to make the behavior diverge widely from the desired goal. Wiener attributes tremors and some other malfunctioning of the human being to improper functioning of negative feedback mechanisms.

Negative feedback can also be used in order to make the large output signal of an amplifier conform closely in shape to the small input. Negative feedback amplifiers were extremely important in communication systems long before the day of cybernetics.

Finally, cybernetics has laid claim to the whole field of automata or complex machines, including telephone switching systems, which have been in existence for many years, and electronic computers, which have been with us only since World War II.

If all this is so, cybernetics includes most of the essence of modern technology, excluding the brute producton and use of power. It includes our knowledge of the organization and function of man as well. Cybernetics almost becomes another word for all of the most intriguing problems of the world. As we have seen, Wiener includes sociological, philosophical, and ethical problems among these.

Thus, even if a man acknowledged being a cyberneticist, that wouldn't give us much of a clue concerning his field of competence, unless he was a universal genius. Certainly, it would not necessarily indicate that he had much knowledge of information theory.

Happily, as I have noted, few scientists would acknowledge themselves as cyberneticists, save perhaps in talking to those whom they regard as hopelessly uninformed. Thus, if cybernetics is overextensive or vague, the overextension or vagueness will do no real harm. Indeed, cybernetics is a very useful word, for it can help to add a little glamor to a person, to a subject, or even to a book. I certainly hope that its presence here will add a little glamor to this one.

CHAPTER XII *Information Theory and Psychology*

I HAVE READ a good deal more about information theory and psychology than I can or care to remember. Much of it was a mere association of new terms with old and vague ideas. Presumably the hope was that a stirring in of new terms would clarify the old ideas by a sort of sympathetic magic.

Some attempted applications of information theory in the field of experimental psychology have, however, been at least reasonably well informed. They have led to experiments which produced valid data. It is hard to draw any conclusions from these data that are both sweeping and certain, but the data do form a basis or at least an excuse for interesting speculations. In this chapter, I propose to discuss some experiments concerning information theory and psychology which are at least down-to-earth enough to grapple with. Naturally I have chosen these largely on the basis of my personal interest and background, but one has to impose some limitations in order to say anything coherent about a broad and less than pellucid field.

It seems to me that an early reaction of psychologists to information theory was that, as entropy is a wonderful and universal measure of amount of information and as human beings make use of information, in some way the difficulty of a task, perhaps the time a man takes to accomplish a set task, must be proportional to the amount of information involved.

This idea is very clearly illustrated in some experiments reported by Ray Hyman, an experimental psychologist, in the *Journal of Experimental Psychology* in 1953. Here I shall describe only one of several of the experiments Hyman made.

A number of lights were placed before a *subject,* as psychologists call an experimentee or laboratory human animal. Each light was labeled with a monosyllabic "name" with which the subject became familiar. After a warning signal, one of the several lights flashed, and the subject thereafter spoke the name of the light as soon as he could. The time interval between the flashing of the light and the speaking of the name was measured.

Sometimes one out of eight lights flashed, the light being chosen at random with equal probabilities. In this case, the information conveyed in enabling the subject to identify the light correctly was log 8, or 3 bits. Sometimes one among 7 light flashed (2.81 bits), sometimes among 6 (2.58 bits), sometimes one out of 5 (2.32 bits), one out of 4 (2.00 bits), one out of 3 (1.58 bits), one out of 2 (1 bit), or one out of 1 (0 bits). The average response time, or *latency,* between the lighting of the light and the speaking of its name was plotted against number of bits, as shown in Figure XII-1.

Clearly, there is a certain latency, or response time, even when only one light is used, the choice among lights is certain, and the information conveyed as to which light is lighted is zero. When more lights are used, the increase in latency is proportional to the information conveyed. Such an increase of latency with the logarithm of the number of alternatives had in fact been noted by a German psychologist, J. Merkel, in 1885. It is certainly a strikingly accurate, reproducible, and a significant aspect of human response.

We note from Figure XII-1 that the increase in latency is about 0.15 second per bit. Some unwary psychologists have jumped to the conclusion that it takes 0.15 second for a human being to respond to 1 bit of information; therefore, the information capacity of a human being is about 1/.15, or about 7 bits per second. Have we discovered a universal constant of human perception or of human thought?

Clearly, in Hyman's experiment the increase in latency is proportional to the uncertainty of the stimulus measured in bits. However, various experiments by various experimenters give

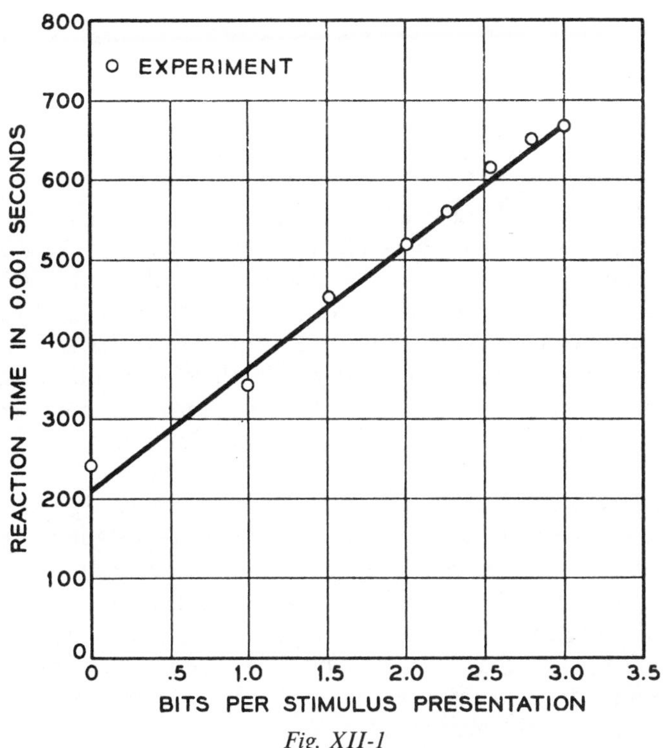

Fig. XII-1

somewhat different rates of increase in seconds per bit. Moreover, data published by G. H. Mowbray and M. V. Rhoades in 1959 show that, after much practice, a subject's performance tends to change so that there is little or no effect of information content on latency. It appears that human beings may have different ways of handling information, a way used in learning, in which number of alternatives is very important, and a way used after much learning, in which number of alternatives, up to a fairly large number, makes little difference. Further, in one sort of experiment, in which a subject depresses one or more keys on which his fingers rest in response to a vibration of the key, it appears that there may be little increase in latency with amount of information right from the start.

Moreover, even if the latency were a constant plus an increment proportional to information content, one could not reasonably assert that this showed that a significant information rate can be obtained by dividing the increased time by the number of bits. We will see that this can lead to fantastic information rates in the sort of experiment which I shall describe next.

H. Quastler made early information-rate experiments in which subjects played random sequences of notes or chords or read lists of randomly chosen words as rapidly as possible, and J. C. R. Licklider did early work on both reading and pointing speed. Before we heard of this work, J. E. Karlin and I embarked on an extensive series of experiments on reading lists of words, which of all experiments gives the highest observed information rate, a rate which is much higher than, for instance, sending Morse code or typing.

Suppose the "sender" of the message chooses an alphabet of, say, 16 words and makes up a list by choosing words among these randomly and with equal probabilities. Then, the amount of choice in designating each word is log 16, or 4 bits. The subject "transmits" the information, translating it into a new form, speech rather than print, by reading the list aloud. If he can read at a rate of 4 words a second, for instance, he transmits information at a rate of 4×4, or 16 bits per second.

Figure XII-2 shows data from three subjects. The words were chosen from the 500 most common words in English. We see that while the reading rate drops somewhat in going from 2 to 4 word vocabularies (or from 1 to 2 bits per word), it is almost constant for vocabularies or alphabets containing from 4 to 256 words (from 2 to 8 bits per word).

Let us now remember the alleged means for getting an information rate from such data as Hyman's, that is, noting the increase in time with increase in bits per stimulus. Consider the dotted average data curve of Figure XII-2. In going from 2 bits per stimulus to 8 bits per stimulus the reading rate doesn't decrease at all; that is, the change in reading time per word is 0, despite an increase of 6 in the number of bits per word. If we divide 6 by 0, we get an information rate of infinity! Of course, this is ridiculous, but it is scarcely more ridiculous than deducing an information

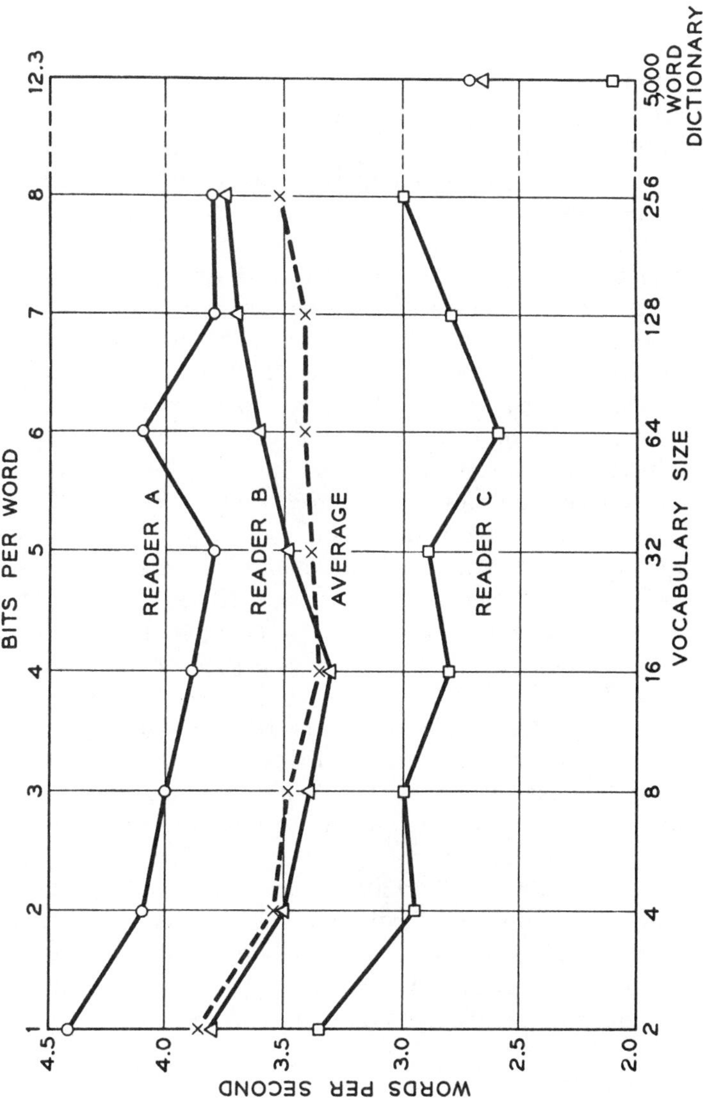
BITS PER WORD
1
2
3
4
5
6
7
8
12.3
WORDS PER SECOND
4.5
4.0
3.5
3.0
2.5
2.0
READER A
READER B
AVERAGE
READER C
VOCABULARY SIZE
2
4
8
16
32
64
128
256
5000 WORD DICTIONARY

Fig. XII-2

rate from such data as Hyman's by dividing increase in number of bits by increase in latency.

Directly from Figure XII-2, we can see that as reader *A* reads 8-bit words at a rate of 3.8 per second, he manages to transmit information at a rate of 8×3.8, or about 30 bits per second. Moreover, when the words put in the list are chosen randomly from a 5,000 word dictionary (12.3 bits per word), he manages to read them at a rate of 2.7 per second, giving a higher information rate of 33 bits per second.

It is clear that no unique information rate can be used to describe the performance of a human being. He can transmit (and, we shall see, respond to or remember) information better under some circumstances than under others. We can best consider him as an information-handling channel or device having certain built-in limitations and properties. He is a very flexible device; he can handle information quite well in a variety of forms, but he handles it best if it is properly encoded, properly adjusted to his capabilities.

What are his capabilities? We see from Figure XII-2 that he is slowed down only a little by increasing complexity. He can read a list of words chosen randomly from an alphabet of 256 about as fast as words chosen from an alphabet of 4. He isn't very speedy compared with machines, and in order to make him perform well we must give him a complex task. This is just what we might have expected.

Complexity eventually does slow him down, however, as we see from the points for an alphabet consisting of all the words in a 5,000 word dictionary. Perhaps there is an optimum alphabet or vocabulary, which has quite a number of bits per word, but not so many words as to slow a man down unduly. Partly to help in finding such a vocabulary, Karlin and I measured reading rate as a function of both number of syllables and "familiarity," that is, whether the word came from the first thousand in order of commonness of occurrence or familiarity, from the tenth thousand, or from the nineteenth thousand. The results are shown in Figure XII-3.

We see that while an increase in number of syllables slows down reading speed, a decrease in familiarity has just as pronounced an effect. Thus, a vocabulary of familiar one syllable words would

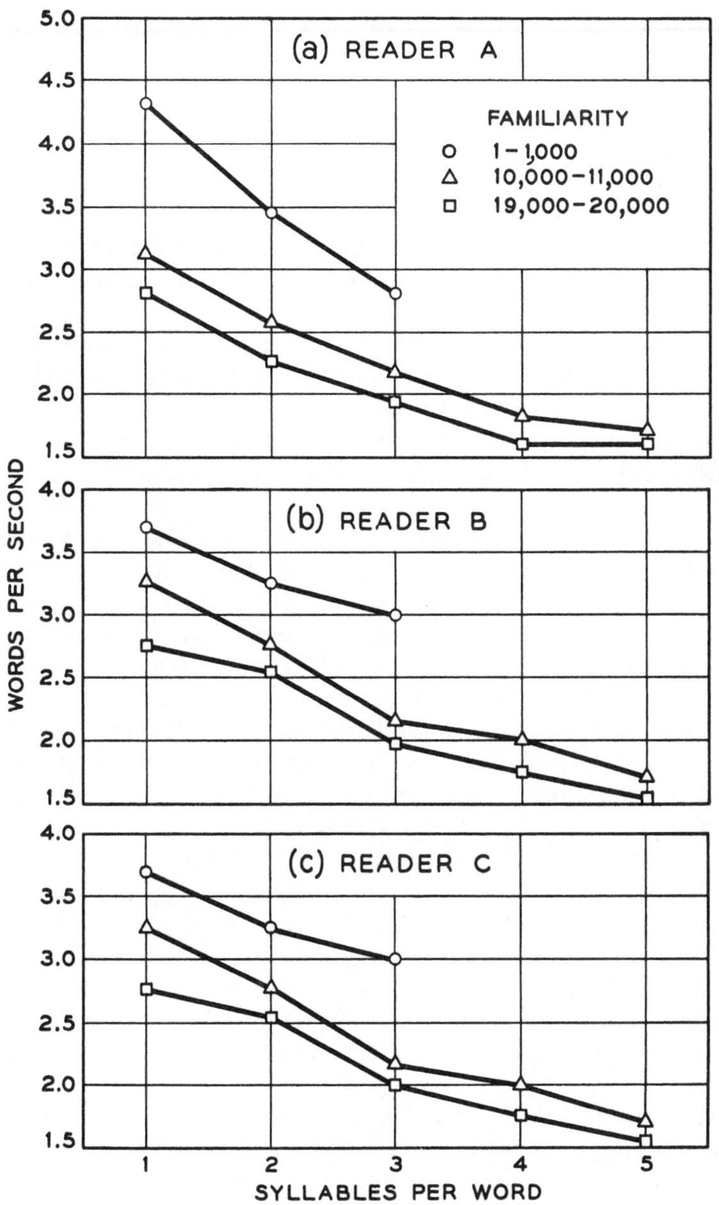
(a) READER A
FAMILIARITY
○ 1-1,000
△ 10,000-11,000
□ 19,000-20,000
(b) READER B
(c) READER C
WORDS PER SECOND
SYLLABLES PER WORD
5.0
4.5
4.0
3.5
3.0
2.5
2.0
1.5
1
2
3
4
5

Fig. XII-3

seem to be a good choice. Using the 2,500 most common monosyllables (2,500 words means 11.3 bits per word) as a "preferred vocabulary," a reader attained a reading speed of 3.7 words per second, giving an information rate of 42 bits per second.

"Scrambled prose," that is, words chosen with the same probabilities as in nontechnical prose but picked at random without grammatical connection, also gave a high information rate. The entropy is about 11.8 bits per word, the highest reading rate was 3.7 words per second, and the corresponding information rate is 44 bits per second.

Perhaps one could gain a little by improving the alphabet, but I don't think one would gain much. At any rate, these experiments gave the highest information rate which has been demonstrated. It is a rate slow by the standards of electrical communication, but it does represent a tremendous number of binary choices—around 2,500 a minute!

What, we may ask, limits the rate? Is it reading through each word letter by letter? In this case the Chinese, who have a single sign for each word, might be better off. But Chinese who read both English and Chinese with facility read randomized lists of common Chinese characters and randomized lists of the equivalent English words at almost exactly the same speed.

Is the limitation a mechanical one? Figure XII-4 shows rates for several tasks. A man can repeat a memorized phrase over twice as fast as he can read randomized words from the preferred list, and he can read prose appreciably faster. It appears that the limitation on reading rate is mental rather than mechanical.

So far, it appears that we cannot characterize a human being by means of a particular information rate. While the difficulty of a task ultimately increases with its information content, the difficulty depends markedly on how well the task is tailored to human abilities. The human being is very flexible in ability, but he has to strain and slow down to do unusual things. And he is quite good at complexity but only fair at speed.

One way of tailoring a task to human abilities is by deliberate, thoughtful experiments. This is analogous to the process of so encoding messages from a message source as to attain the highest possible rate of information transmission over a noisy channel.

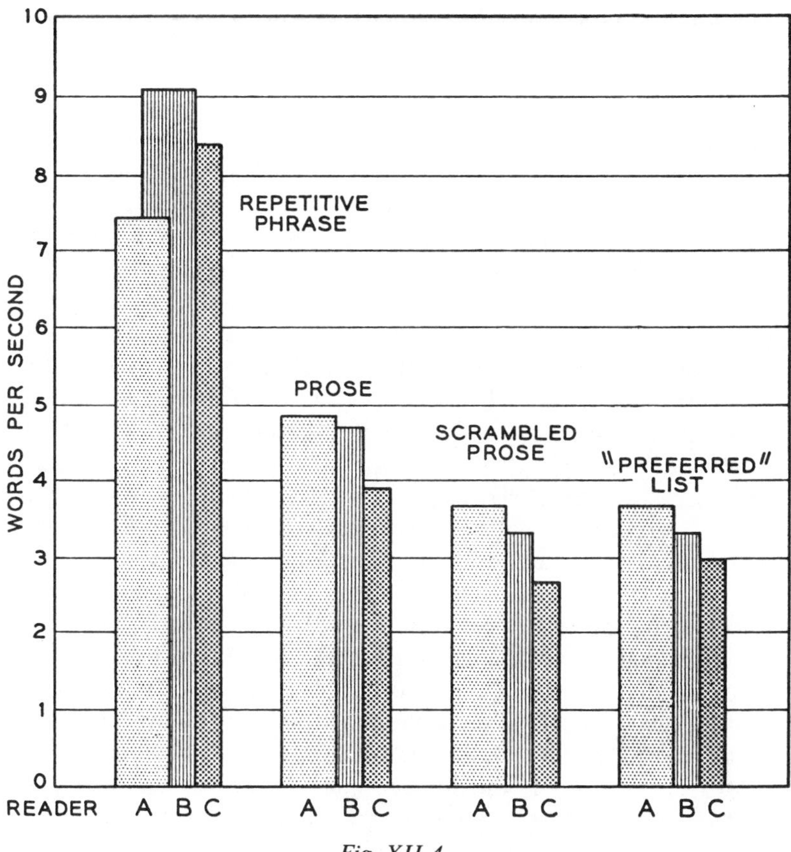

Fig. XII-4

This was discussed in Chapter VIII, and the highest attainable rate was called the channel capacity. The "preferred list" of the 2,500 most frequently used monosyllables was devised in a deliberate effort to attain a high information rate in reading aloud randomized lists of words.

We may note, however, that choosing words at random with the probabilities of their occurrence in English text gives as high or a little higher information rate. Have the words of the English language and their frequencies of occurrence been in some way fitted to human abilities by a long process of unconscious experiment and evolution?

We have seen in Chapter V that the probability of occurrence of a word in English text is very nearly inversely proportional to its rank. That is, the hundredth most common word occurs about a hundredth as frequently as the most common word, and so on. Figure V-2 illustrates this relation, which was first pointed out by George Kingsley Zipf, who ascribed it to a principle of least effort.

Clearly, Zipf's law cannot be entirely correct in this simple form. We saw in Chapter V that word probabilities cannot be inversely proportional to the rank of the word for all words; if they were, the sum of the probabilities of all words would be greater than unity. There have been various attempts to modify, derive, and explain Zipf's law, and we will discuss these somewhat later. However, we will at first regard Zipf's law in its original and simplest form as an *approximate* description of an aspect of human behavior in generating language, a description which Zipf arrived at empirically by examining the statistics of actual text.

Zipf, as we have noted, associated his law with a principle of least effort. Attempts have been made to identify the effort or "cost" of producing text with the number of letters in text. However, most linguists regard language primarily as the spoken language, and it seems unlikely that speaking, reading, or writing habits are dictated primarily by the numbers of letters used in words.

In fact, we noted in the information-rate experiments which we just considered that reading rates are about the same for common Chinese ideographs and for the equivalent words in English written out alphabetically. Further, we have noted from Figure XII-3 that commonness or familiarity has an influence on reading time as great as does number of syllables.

Could we not, perhaps, take reading time as a measure of effort? We might think, for instance, that common words are more easily accessible to us, that they can be recognized or called forth with less effort or cost than uncommon words. Perhaps the human brain is so organized that a few words can be stored in it in such a fashion that they can be recognized and called forth easily and that many more can be stored in a fashion which makes their use less easy. We might believe that reading time is a measure of accessibility, ease of use, of cost.

We might imagine, further, that in using language, human beings choose words in such a way as to transmit as much information as possible for a given cost. If we identify cost with time of utterance, we would then say that human beings choose words in such a way as to convey as much information as possible in a given time of speaking or in a given time of reading aloud.

It is an easy mathematical task to show that if a speaking time t_r is associated with the rth word in order of commonness, then for a message composed of randomly chosen words the information rate will be greatest if the rth word is chosen with a probability $p(r)$ given by

$$p(r) = 2^{-\mathrm{c}t_r} \tag{12.1}$$

Here c is a constant chosen to make the sum of the probabilities for all words add up to unity. This mathematical relation says that words with a long reading time will be used less frequently than words with a short reading time, and it gives the exact relation which must hold if the information rate is to be maximized.

Now, if Zipf's law holds, the probability of occurrence of the rth word in order of commonness must be given by

$$p(r) = \frac{\mathrm{A}}{r} \tag{12.2}$$

Here A is another constant. Thus, from 12.1 and 12.2 we must have

$$\frac{\mathrm{A}}{r} = 2^{-\mathrm{c}t_r} \tag{12.3}$$

By using a relation given in the Appendix, this relation can be re-expressed

$$t_r = \mathrm{a} + \mathrm{b} \log r \tag{12.4}$$

Here a and b are constants which must be determined by examining the relation of the reading time t_r and the order of commonness or rank of a word, r. If Zipf's law is true and if the information rate is maximized for words chosen randomly and independently with probabilities given by Zipf's law, then relation 12.4 should hold for experimental data.

Of course, words aren't chosen randomly and independently in constructing English text, and hence we cannot say that word probabilities in accord with relation 12.1 actually *would* maximize information transmission per unit time. Nonetheless, it would be interesting to know whether predictions based on a random and independent choice of words do hold for the reading of actual English text.

Benoit Mandelbrot, a mathematician much interested in linguistic problems, has considered this matter in connection with reading-time data taken by D. H. Howes, an experimental psychologist. R.R. Riesz, an experienced experimenter in the field of psychophysics, and I have also attempted to compare equation 12.4 with human behavior.

There is a difficulty in making such a comparison. It seems fairly clear that reading speed is limited by word *recognition,* not by word *utterance.* A man may be uttering a long familiar word while he is recognizing a short, unfamiliar word. To get around this difficulty it seemed best to do some averaging by measuring the total time of utterance for three successive words and then comparing this with the sum of the times for the words computed by means of 12.4.

Riesz ingeniously and effectively did this and obtained the data of Figure XII-5. In the test, a subject read a paragraph as fast as possible. Certainly, a straight line according to 12.4 fits the data as well as any curve would. But the points are too scattered to prove that 12.4 really holds.

Moreover, we should expect such a scatter, for the rank *r* corresponds to commonness of occurrence in prose from a variety of sources, but we have used it as indicating the subject's experience with and familiarity with the word. Also, as we see from Figure XII-3, word length may be expected to have some effect on reading time. Finally, we have disregarded relations among successive words.

This sort of experiment is extremely exasperating. One can see other experiments which he might do, but they would be time consuming, and there seems little chance that they would establish anything of general significance in a clear-cut way. Perhaps a genius will unravel the situation some day, but the wary psychologist is more apt to seek a field in which his work promises a definite, unequivocal outcome.

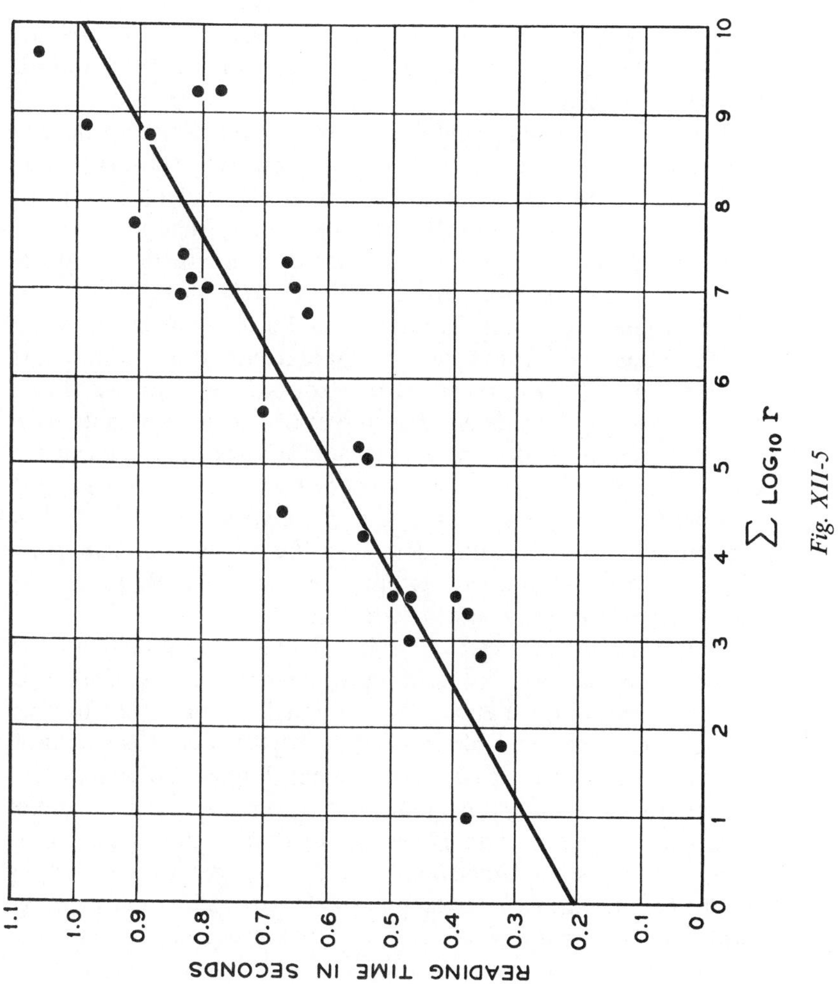
READING TIME IN SECONDS
1.1
1.0
0.9
0.8
0.7
0.6
0.5
0.4
0.3
0.2
0.1
0
0
1
2
3
4
5
6
7
8
9
10
$\sum \text{LOG}_{10}\, r$

Fig. XII-5

The foregoing work does at least suggest that word usage *may* be governed by economy of effort and that economy of effort *may* be measured as economy of time. We still wonder, however, whether this is the outcome of a trained ability to cope with the English language or whether language somehow becomes adapted to the mental abilities of people. What about the number of words we use, for instance?

People sometimes measure the vocabulary of a writer by the total number of different words in his works and the vocabulary of an individual by the number of different words he understands. However, rare and unusual words make up a small fraction of spoken or written English. What about the words that constitute most of language? How numerous are these?

One might assert that the number of words used should reflect the complexity of life and that we would need more in Manhattan than in Thule (before the Air Force, of course). But, we always have the choice of using either different words or combinations of common words to designate particular things. Thus, I can say either "the blonde girl," "the redheaded girl," "the brunette girl" or I can say "the girl with light hair," "the girl with red hair," "the girl with dark hair." In the latter case, the words *with, light, dark, red,* and *hair* serve many other purposes, while *blonde, redheaded,* and *brunette* are specialized by contrast.

Thus, we could construct an artificial language with either fewer or more common words than English has, and we could use it to say the same things that we say in English. In fact, we can if we wish regard the English alphabet of twenty six letters as a reduced language into which we can translate any English utterance.

Perhaps, however, *all* languages tend to assume a basic size of vocabulary which is dictated by the capabilities and organization of the human brain rather than by the seeming complexity of the environment. To this basic language, clever and adaptable people can, of course, add as many special and infrequently used words as they desire to or can remember.

Zipf has studied just this matter by means of the graphs illustrating his law. Figure XII-6 [1] shows frequency (number of times a

[1] Reproduced from George Kingsley Zipf, *Human Behavior and the Principle of Least Effort,* Addison-Wesley Publishing Company, Reading, Mass., 1949.

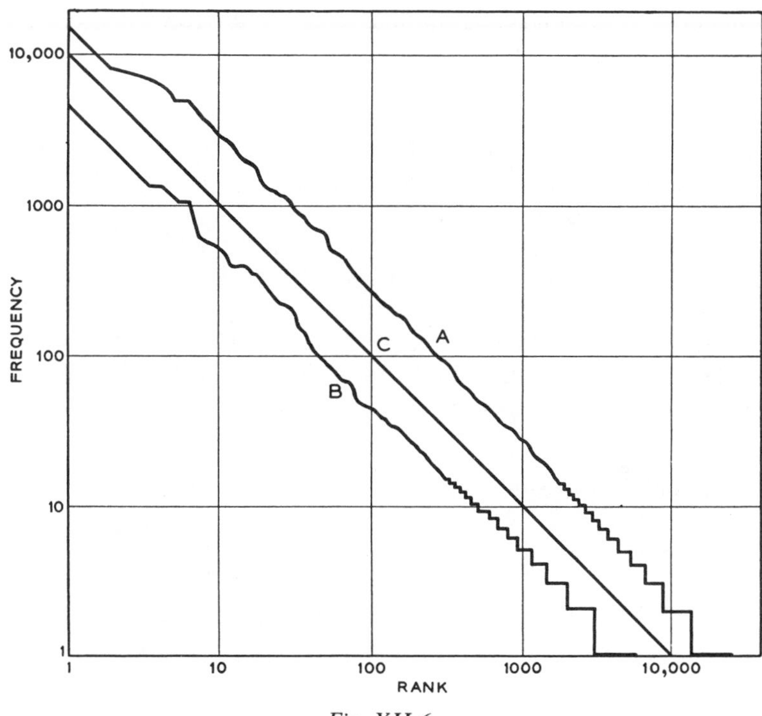

Fig. XII-6

word is used) plotted against rank (order of commonness) for 260,430 running words of James Joyce's *Ulysses* (curve *A*) and for 43,989 running words from newspapers (curve *B*). The straight line *C* illustrates Zipf's idealized curve or "law."

Clearly, the heights of *A* and *B* are determined merely by the number of words in the sample; the slope of the curve and its constancy with length of sample are the important things. The steps at the lower right of the curves, of course, reflect the fact that infrequent words can occur once, twice, thrice, and so on in the sample but not 1.5 or 2.67 times.

When we idealize such curves to a 45° line, as in curve *C,* we note that more is involved than the mere slope of the line. We start our frequency measurement with words which occur only once; that is, the lower right hand corner of the graph represents a frequency of occurrence of 1. Similarly, the rank scale starts with 1, the rank assigned to the most frequently used word. Thus, vertical and horizontal scales start as 1, and equal distances along them were chosen in the first place to represent equal increases in number. We see that the 45° Zipf-law line tells us that *the number of different words in the sample* must equal *the number of occurrences of the most frequently used word.*

We can go further and say that if Zipf's law holds in this strict and primitive form, a number of words equal to the square root of the number of *different* words in the passage will make up *half* of all the words in the sample. In Figure XII-7 the number N of different words and the number V of words constituting half the passage are plotted against the number L of words in the passage.

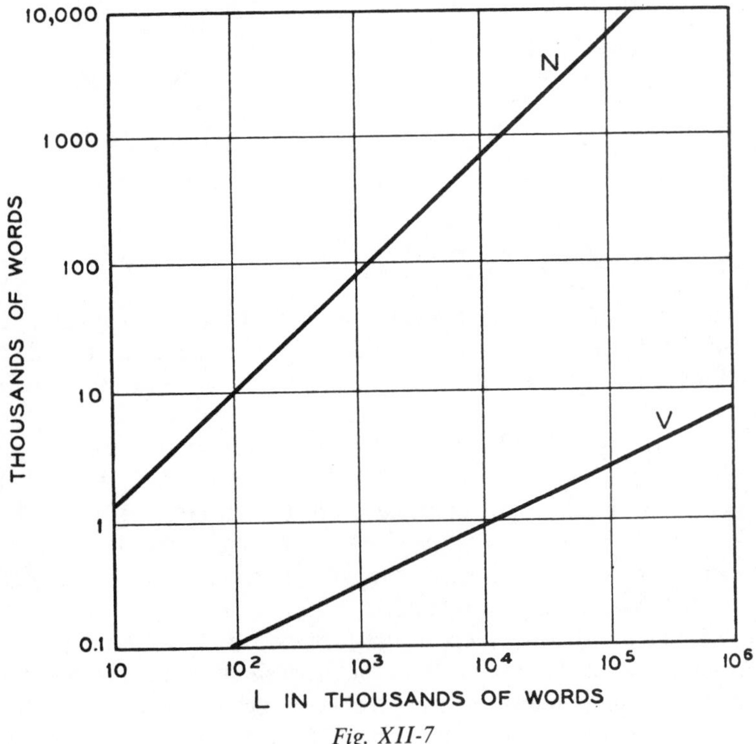

Fig. XII-7

In his second derivation of a modified form of Zipf's law as a consequence of certain initial assumptions, Mandelbrot assumes that word frequencies are such as to maximize the information for a given cost. As a simple case, he assumes that each letter has a particular cost and that the cost of each word (that is, of each sequence of letters ending in a space) is the sum of the costs of its letters. This leads him to the same expression as the other derivation, that is, to equation 12.5. The interpretation of the different symbols is different, however. The constant B can be less than unity if the total number of allowable words is finite.

Regardless of the meaning of the constants, P, V, and B in expression 12.5, we can, if we wish, merely give them such values as will make the curve defined by 12.5 best fit statistical data derived from actual text. Certainly, we can fit actual data better his way than we can if we assume that $V = 0$ and $B = 1$ (corresponding to Zipf's original law). In fact, by so choosing the values of P, V, and B, equation 12.5 can be made to fit available data very well in all but a few exception cases. In the cases of modern Hebrew of around 1930 and Pennsylvania Dutch, which is a mixture of languages, a value of B smaller than 1 gives the best fit.

According to Mandelbrot, the wealth of vocabulary is measured chiefly by the value of B; if B is much greater than 1, a few words are used over and over again; if B is nearer to 1, a greater variety of words is used. Mandelbrot observes that as a child grows, B decreases from values around 1.6 to values around 1.15 or to a value around 1 if the child happens to be James Joyce.

Certainly, equation 12.5 fits data better than Zipf's original law does. It overcomes the objection that, according to Zipf's original law, the probability of the word *the* should depend on the length of the sample of text. This does not mean, however, that Mandelbrot's explanation or derivation of equation 12.5 is necessarily correct. Further, it is possible that some other mathematical expression would fit data concerning actual text even better. A much more thorough study would be necessary to settle such questions.

Zipf's law holds for many other data than those concerning word usage. For instance, in most countries it holds for population of cities plotted against rank in size. Thus, the tenth largest city has about a tenth the population of the largest city, and so on.

However, the fact that the law holds in different cases may be fortuitous. The inverse-square law holds for gravitational attraction and also for intensity of light at different distances from the sun, yet these two instances of the law cannot be derived from any common theory.

It is clear that our ability to receive and handle information is influenced by inherent limitations of the human nervous system. George A. Miller's law of 7 plus-or-minus 2 is an example. This states that after a short period of observation, a person can remember and repeat the names of from 5 to 9 simple, familiar objects, such as binary or decimal digits, letters, or familiar words.

By means of a tachistoscope, a brightly illuminated picture can be shown to a human subject for a very short time. If he is shown a number of black beans, he can give the number correctly up to perhaps as many as 9 beans. Thus, one flash can convey a number 0 through 9, or 10 possibilities in all. The information conveyed is log 10, or 3.3 bits.

If the subject is shown a sequence of binary digits, he can recall correctly perhaps as many as 7, so that 7 bits of information are conveyed.

If the subject is shown letters, he can remember perhaps 4 or 5, so that the information is as much as 5 log 26 bits, or 23 bits.

The subject can remember perhaps 3 or 4 short, common words, somewhat fewer than 7 − 2. If these are chosen from the 500 most common words, the information is 3 log 500, or 27 bits.

As in the case of the reading rate experiments, the gain due to greater complexity outweighs the loss due to fewer items, and the information increases with increasing complexity.

Now, both Miller's 7 plus-or-minus-2 rule and the reading rate experiments have embarrassing implications. If a man gets only 27 bits of information from a picture, can we transmit by means of 27 bits of information a picture which, when flashed on a screen, will satisfactorily imitate any picture? If a man can transmit only about 40 bits of information per second, as the reading rate experiments indicate, can we transmit TV or voice of satisfactory quality using only 40 bits per second?

In each case I believe the answer to be no. What is wrong? What is wrong is that we have measured what gets *out* of the human being, not what goes *in*. Perhaps a human being can in some sense only notice 40 bits a second worth of information, but he has a choice as to what he notices. He might, for instance, notice the girl or he might notice the dress. Perhaps he notices more, but it gets away from him before he can describe it.

Two psychologists, E. Averback and G. Sperling, studied this problem in similar manners. Each projected a large number (16 or 18) of letters tachistoscopically. A fraction of a second later they gave the subject a signal by means of a pointer or tone which indicated which of the letters he should report. If he could unfailingly report any indicated letter, all the letters must have "gotten in," since the letter which was indicated was chosen randomly.

The results of these experiments seem to show that far more than 7 plus-or-minus-2 items are seen and stored briefly in the organism, for a few tenths of a second. It appears that 7 plus-or-minus-2 of these items can be transferred to a more permanent memory at a rate of about one item each hundredth of a second, or less than a tenth of a second for all items. This other memory can retain the transferred items for several seconds. It appears that it is the size limitation of this longer-term memory which gives us the 7 plus-or-minus-2 figure of Miller.

Human behavior and human thought are fascinating, and one could go on and on in seeking relations between information theory and psychology. I have discussed only a few selected aspects of a broad field. One can still ask, however, is information theory really highly important in psychology, or does it merely give us another way of organizing data that might as well have been handled in some other manner? I myself think that information theory has provided psychologists with a new and important picture of the process of communication and with a new and important measure of the complexity of a task. It has also been important in stirring psychologists up, in making them re-evaluate old data and seek new data. It seems to me, however, that while information theory provides a central, universal structure and organization for electrical communication, it constitutes only an attractive area in psychology. It also adds a few new and sparkling expressions to the vocabulary of workers in other areas.

CHAPTER XIII *Information Theory and Art*

SOME YEARS AGO when a competent modern composer and professor of music visited the Bell Laboratories, he was full of the news that musical sounds and, in fact, whole musical compositions can be reduced to a series of numbers. This was of course old stuff to us. By using pulse code modulation, one can represent any electric or acoustic wave form by means of a sequence of sample amplitudes.

We had considered something that the composer didn't appreciate. In order to represent fairly high-quality music, with a band width of 15,000 cycles per second, one must use 30,000 samples per second, and each one of these must be specified to an accuracy of perhaps one part in a thousand. We can do this by using three decimal digits (or about ten binary digits) to designate the amplitude of each sample.

A composer could exercise complete freedom of choice among sounds simply by specifying a sequence of 30,000 three-digit decimal numbers a second. This would allow him to choose from among a number of twenty-minute compositions which can be written as 1 followed by 108 million 0's—an inconceivably large number. Putting it another way, the choice he could exercise in composing would be 300,000 bits per second.

Here we sense what is wrong. We have noted that by the fastest demonstrated means, that is, by reading lists of words as rapidly as possible, a human being demonstrates an information rate of no more than 40 bits per second. This is scarcely more than a ten-thousandth of the rate we have allowed our composer.

Further, it may be that a human being can make use of, can appreciate, information only at some rate even less than 40 bits per second. When we listen to an actor, we hear highly redundant English uttered at a rather moderate speed.

The flexibility and freedom that a composer has in expressing a composition as a sequence of sample amplitudes is largely wasted. They allow him to produce a host of "compositions" which to any human auditor will sound indistinguishable and uninteresting. Mathematically, white Gaussian noise, which contains all frequencies equally, is the epitome of the various and unexpected. It is the least predictable, the most original of sounds. To a human being, however, all white Gaussian noise sounds alike. Its subtleties are hidden from him, and he says that it is dull and monotonous.

If a human being finds monotonous that which is mathematically most various and unpredictable, what does he find fresh and interesting? To be able to call a thing new, he must be able to distinguish it from that which is old. To be distinguishable, sounds must be to a degree familiar.

We can tell our friends apart, we can appreciate their particular individual qualities, but we find much less that is distinctive in strangers. We can, of course, tell a Chinese from our Caucasian friends, but this does not enable us to enjoy variety among Chinese. To do that we have to learn to know and distinguish among many Chinese. In the same way, we can distinguish Gaussian noise from Romantic music, but this gives us little scope for variety, because all Gaussian noise sounds alike to us.

Indeed, to many who love and distinguish among Romantic composers, most eighteenth-century music sounds pretty much alike. And to them Grieg's *Holberg Suite* may sound like eighteenth-century music, which it resembles only superficially. Even to those familiar with eighteenth-century music, the choral music of the sixteenth century may seem monotonous and undistinguishable. I know, too, that this works in reverse order, for some partisans of Mozart find Verdi monotonous, and to those for whom Verdi affords tremendous variety much modern music sounds like undistinguishable noise.

Of course a composer wants to be free and original, but he also wants to be known and appreciated. If his audience can't tell one of his compositions from another, they certainly won't buy recordings of many different compositions. If they can't tell his compositions from those of a whole school of composers, they may be satisfied to let one recording stand for the lot.

How, then, can a composer make his compositions distinctive to an audience? Only by keeping their entropy, their information rate, their variety within the bounds of human ability to make distinctions. Only when he doles his variety out at a rate of a very few bits per second can he expect an audience to recognize and appreciate it.

Does this mean that the calculating composer, the information-theoretic composer so to speak, will produce a simple and slow succession of randomly chosen notes? Of course not, not any more than a writer produces a random sequence of letters. Rather, the composer will make up his composition of larger units which are already familiar in some degree to listeners through the training they have received in listening to other compositions. These units will be ordered so that, to a degree, a listener expects what comes next and isn't continually thrown off the track. Perhaps the composer will surprise the listener a bit from time to time, but he won't try to do this continually. To a degree, too, the composer will introduce entirely new material sparingly. He will familiarize the listener with this new material and then repeat the material in somewhat altered forms.

To use the analogy of language, the composer will write in a language which the listener knows. He will produce a well-ordered sequence of musical words in a musically grammatical order. The words may be recognizable chords, scales, themes, or ornaments. They will succeed one another in the equivalents of sentences or stanzas, usually with a good deal of repetition. They will be uttered by he familiar voices of the orchestra. If he is a good composer, he will in some way convey a distinct and personal impression to the skilled listener. If he is at least a skillful composer, his composition will be intelligible and agreeable.

Of course, none of this is new. Those quite unfamiliar with information theory could have said it, and they have said it in other words. It does seem to me, however, that these facts are particularly pertinent to a day in which composers, and other artists as well, are faced with a multitude of technical resources which are tempting, exasperating, and a little frightening.

Their first temptation is certainly to choose too freely and too widely. M. V. Mathews of the Bell Laboratories was intrigued by the fact that an electronic computer can create any desired wave form in response to a sequence of commands punched into cards. He devised a program such that he could specify one note by each card as to wave form, duration, pitch, and loudness. Delighted with the freedom this afforded him, he had the computer reproduce rapid rhythmic passages of almost unplayable combinations, such as three notes against four with unusual patterns of accent. These ingenious exercises sounded, simply, chaotic.

Very skillful composers, such as Varèse, can evoke an impression of form and sense by patching together all sorts of recorded and modified sounds after the fashion of *musique concrète.* Many appealing compositions utilizing electronically generated sounds have already been produced. Still, the composer is faced with difficulties when he abandons traditional resources.

The composer can choose to make his compositions much simpler than he would if he were writing more conventionally, in order not to lose his audience. Or he and others can try to educate an audience to remember and distinguish among the new resources of which they avail themselves. Or the composer can choose to remain unintelligible and await vindication from posterity. Perhaps there are other alternatives; certainly there are if the composer has real genius.

Does information theory have anything concrete to offer concerning the arts? I think that it has very little of serious value to offer except a point of view, but I believe that the point of view may be worth exploring in the brief remainder of this chapter.

In Chapters III, VI, and XII we considered language. Language consists of an alphabet or vocabulary of words and of grammatical rules or constraints concerning the use of words in grammatical text. We learned to distinguish between the features of text which are dictated by the vocabulary and the rules of grammar and the actual choice exercised by the writer or speaker. It is only this element of choice which contributes to the average amount of information per word. We saw that Shannon has estimated this to be between 3.3 and 7.2 bits per word. It must also be this choice which enables a writer or speaker to convey meaning, whatever that may be.

The vocabulary of a language is large, although we have seen in Chapter XII that a comparatively few words make up the bulk of any text. The rules of grammar are so complicated that they have not been completely formulated. Nonetheless, most people have a large vocabulary, and they know the rules of grammar in the sense that they can recognize and write grammatical English.

It is reasonable to assume a similarly surprisingly large knowledge of musical elements and of relations among them on the part of the person who listens to music frequently, attentively, and appreciatively. Of course, it is not necessary that the listener be able to formulate his knowledge for him to have it, any more than the writer of grammatical English need be able to formulate the rules of English grammar. He need not even be able to write music according to the rules, any more than a mute who understands speech can speak. He can still in some sense *know* the rules and make use of his knowledge in listening to music.

Such a knowledge of the elements and rules of the music of a particular nation, era, or school is what I have referred to as "knowing the language of music" or of a style of music. However much the rules of music may or may not be based on physical laws, a knowledge of a language of music must be acquired by years of practice, just as the knowledge of a spoken language is. It is only by means of such a knowledge that we can distinguish the style and individuality of a composition, whether literary or musical. To the untutored ear, the sounds of music will seem to be examples chosen not from a restricted class of learned sounds but from all the infinity of possible sounds. To the untutored ear, the mechanical workings of the rules of music will seem to represent choice and variety. Thus, the apparent complexity of music will overwhelm the untutored auditor or the auditor familiar only with some other language of music.

We should note that we can write sense while violating the rules of grammar to a degree (me heap big injun). We might liken the intelligibility of this sentence to an English-speaking person to our ability to appreciate music which is somewhat strange but not entirely foreign to our experience. We should also note that we can write nonsense while obeying the rules of grammar carefully (the alabaster word spoke silently to the purple). It is to this second possibility to which I wish to address myself in a moment. I will remark first, however, that while one can of course both write sense and obey the rules while doing so, he often exposes his inadequacies to the public gaze by thus being intelligible.

It is no news that we can dispense with sense almost entirely while retaining a conventional vocabulary and some or many rules. Thus, Mozart provided posterity with a collection of assorted, numbered bars in 3/8 time, together with a set of rules (Koechel 294D). By throwing dice to obtain a sequence of random numbers and choosing successive bars by means of the rules, even the nonmusical amateur can "compose" an almost endless number of little waltzes which sound like somewhat disorganized Mozart. An example is shown in Figure XIII-1. Joseph Haydn, Maximilian Stadler, and Karl Philipp Emanuel Bach are said to have produced similar random music. In more recent times, John Cage has used random processes in the choice of sequences of notes.

In ignorance of these illustrious predecessors, in 1949 M. E. Shannon (Claude Shannon's wife) and I undertook the composition of some very primitive statistical or stochastic music. First we made a catalog of allowed chords on roots I–VI in the key of C. Actually, the catalog covered only root I chords; the others were

Fig. XIII-1

derived from these by rules. By throwing three specially made dice and by using a table of random numbers, a number of compositions were produced.

In these compositions, the only rule of chord connection was that two succeeding chords have a common tone in the same voice. This let the other voices jump around in a wild and rather unsatisfactory manner. It would correspond to the use of a simple and consistent but incorrect digram probability in the construction of synthetic text, as illustrated in Chapter III.

While the short-range structure of these compositions was very primitive, an effort was made to give them a plausible and reasonably memorable, longer-range structure. Thus, each composition consisted of eight measures of four quarter notes each. The long-range structure was attained by making measures 5 and 6 repeat measures 1 and 2, while measures 3 and 4 differed from measures 7 and 8. Thus, the compositions were primitive rondos. Further, it was specified that chords 1, 16, and 32 have root I and chords 15 and 31 have either root IV or root V, in order to give the effect of a cadence.

Although the compositions are formally rondos, they resemble hymns. I have reproduced one as Figure XIII-2. As all hymns should have titles and words, I have provided these by nonrandom means. The other compositions sound much like the one given. Clearly, they are all by the same composer. Still, after a few hearings they can be recognized as different. I have even managed to grow fond of them through hearing them too often. They must grate on the ears of an uncorrupted musician.

In 1951, David Slepian, an information theorist of whom we have heard before, took another tack. Following some early work by Shannon, he evoked such statistical knowledge of music as lay latent in the breasts of musically untrained mathematicians who were near at hand. He showed such a subject a quarter bar, a half bar, or three half bars of a "composition" and asked the subject to add a sensible succeeding half bar. He then showed another subject an equal portion including that added half bar and got another half bar, and so on. He told the subjects the intended styles of the compositions.

Fig. XIII-2

In Figure XIII-3, I show two samples: a fragment of a chorale in which each half bar was added on the basis of the preceding half bar and a fragment of a "romantic composition," in which each half bar was added on the basis of the preceding three half bars. It seems to me surprising that these "compositions" hang together as well as they do, despite the inappropriate and inadmissible chords and chord sequences which appear. The distinctness

Fig. XIII-3

of the styles is also arresting; apparently the mathematicians had quite different ideas of what was appropriate in a chorale and what was appropriate in a romantic composition.

Slepian's experiment shows the remarkable flexibility of the human being as well as some of his fallibility. True stochastic processes are apt to be more consistent but duller. A number have been used in the composition of music.

There is no doubt that a computer supplied with adequate statistics describing the style of a composer could produce random music with a recognizable similarity to a composer's style. The nursery-tune style demonstrated by Pinkerton and the diversity of styles evoked by Hiller and Isaacson, which I will describe presently, illustrate this possibility.

In 1956, Richard C. Pinkerton published in the *Scientific American* some simple schemes for writing tunes. He showed how a note could be chosen on the basis of its probability of following the particular preceding note and how the probabilities changed with respect to the position of the note in the bar. Using probabilities derived from nursery tunes, he computed the entropy per note, which he found to be 2.8 bits. I feel sure that this is quite a bit too high, because only digram probabilities were considered. He also presented a simple finite-state machine which could be used to generate banal tunes, much as the machine of Figure III-1 generates "sentences."

In 1957, F. B. Brooks, Jr., A. L. Hopkins, Jr., P. G. Neumann, and W. V. Wright published an account of the statistical composition of music on the basis of an extensive statistical study of hymn tunes.

In 1956, the Burroughs Corporation announced that they had used a computer to generate music, and, in 1957, it was announced that Dr. Martin Klein and Dr. Douglas Bolitho had used the Datatron computer to write "popular" melodies. Jack Owens set words to one, and it was played over the ABC network as *Push Button Bertha*. No doubt many others have done similar things.

It remained, however, for L. A. Hiller, Jr., and L. M. Isaacson of the University of Illinois to make a really serious experiment with computer music. Hiller and Isaacson succeeded in formulating the rules of four-part, first-species counterpoint in such a way that a computer could choose notes randomly and reject them if they violated the rules.

Because the rules involve, except in connection with the concluding cadence, only direct relations among three successive notes, the music tends to wander, but over a short range it sounds surprisingly good. A sample is shown in Figure XIII-4.[1]

Hiller and Isaacson went on to demonstrate that they could use the computer to generate interesting rhythmic and dynamic patterns and to generate "Markoff-chain" music, in which successive note selection depended on probability functions computed from tables derived from various considerations of overtones or harmonics. In this case they generated a coda according to a simple prescription.

As it stands, this music, which was brought together and published as the *Illiac Suite for String Quartet,* has a good deal of local structure but is weak and wandering as a whole. The imposition

[1] Reproduced from L. A. Hiller, Jr., and L. M. Isaacson, *Illiac Suite for String Quartet,* New Music, 1957, by permission of Theodore Presser Company, Bryn Mawr, Pa.

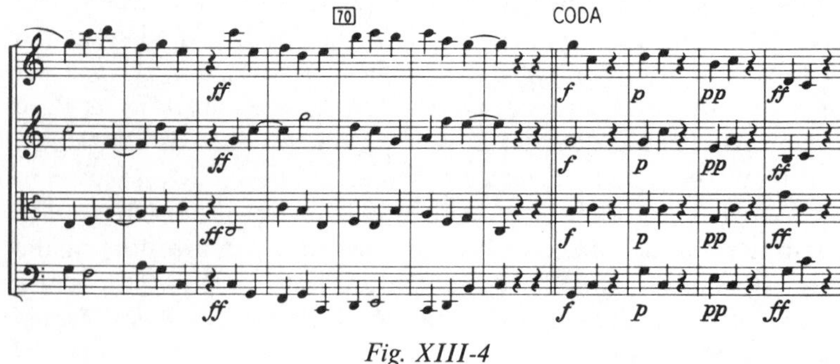

Fig. XIII-4

of some simple pattern or repetition might have helped considerably. This could be of a strictly deterministic nature, as in the case of the prescribed repetitions in a rondo, or it could be of the nature of Chomsky's grammar, which we have considered in Chapter VI. It is clear, however, that it is foolish to try to attain long-range structure simply by relating a note to the immediately preceding notes by digram, trigram, and higher probabilities. The relation must be among *parts* of the composition, not simply among notes.

The work of Hiller and Isaacson does demonstrate conclusively that a computer can take over many musical chores which only human beings had been able to do before. A composer, and especially an unskilled composer, might very well rely on a computer for much routine musical drudgery. The composer could merely guide the main pattern of the composition and let the computer fill in details of harmony and counterpoint, according to a specification of style or period. Further, the computer could be used to try out proposed new rules of composition, such as new rules of counterpoint or harmony, with whose use and consequences the composer might have little experience and familiarity.

In these days we hear that cybernetics will soon give us machines which learn. If they learn in a complicated enough sense of the word, why couldn't they learn what we like, even when we don't know ourselves? Thus, by rewarding or punishing a computer for the success or failure of its efforts, we might so condition the computer that when we pressed a button marked Spanish, classical, rock-and-roll, sweet, etc., it would produce just what we wanted in connection with the terms. Such thoughts are intriguing, but they are of course nonsense in our day and will probably remain so for a long time to come.

Music is not all of art. I began with music because it offers an apt means for illustrating in an unusual context some ideas derived from information theory. We could just as well draw our illustrations from the use of language. Indeed, experiments with the stochastic production of text have been perhaps more widely cultivated than experiments with music.

A professor at the Grand Academy of Lagoda showed Captain Lemuel Gulliver a word frame consisting of lettered blocks mounted on shafts. The professor turned these at random and sought new wisdom in the patterns of letters which appeared.

Here we see just the wrong application of a stochastic process in the generation of text. Certainly, this will not give us new knowledge. Who would take the uncorroborated word of a random process? There are all too many unsubstantiated statements available; what we need to know is what is so and what isn't.

Nonetheless, a stochastic process can produce some interesting effects. In Chapter III we noted Shannon's approximations to English text. These were made by using letter digram and trigram frequencies and a table of random numbers. We have seen that they contain some interesting "words."

To me, *deamy* has a pleasant sound; I would take "it's a deamy idea" in a complimentary sense. On the other hand, I'd hate to be denounced as *ilonasive.* I would not like to be called *grocid;* perhaps it reminds me of gross, groceries, and gravid. *Pondenome,* whatever it may be, is at least dignified.

I repeat Shannon's second-order word approximation here:

> The head and in frontal attack on an English writer that the character of this point is therefore another method for the letters that the time who ever told the problem for an unexpected.

I find this disquieting. I feel that the English writer is in mortal peril, yet I cannot come to his aid because the latter part of the message is garbled.

In seeking less garbled material, as I noted in Chapter VI, I wrote three grammatically connected words in a column from the top down on a slip of paper. I showed them to a friend, asked him to make up a sentence in which they occurred, and then to add the next word in this sentence. I then folded over the top word of the four I now had and showed the visible three to another friend and got another word from him. After canvassing twenty friends, I had the following:

> When morning broke after an orgy of wild abandon he said her head shook vertically aligned in a sequence of words signifying what . . .

Later examples are:

> One day when I went to what was Dionysus thinking of women without men go off half way decent impression . . .
>
> I forget whether he went on and on. Finally he stipulated that this must stop immediately after this. The last time I saw him when she lived. It happened one frosty look of trees waving gracefully against the wall. You never can . . .

We see that a seeming order of meaning persists over groups of far more than four words. Eventually, however, the text wanders. The long-range wandering is of course attributable to the fact that there is no long-range, persistent purpose or meaning guiding the choice of words. We sometimes see a similar quality in the utterances of schizophrenics. W. H. Hudson illustrates a more gradual wandering admirably by a character in his *The Purple Land:* Uncle Anselmo never gets to the end of a story because he continually wanders on to new ground.

One can add a certain amount of long-range order by writing, in view, at the bottom of the slip of paper to which people add words a title which indicates what the passage is supposed to be about.

Dr. Donald A. Dunn of the Stanford Electronics Laboratory has kindly supplied me with some examples in which the person adding a word saw only one preceding word, together with the title:

Men and Women

Eve loved intensely sentimental or not sufficient tonight wherever you may die before yesterday again and whatever m'love misbehaves.

The seduced are compatible unusual family life seemed wonderful experience for tenderness forever yours.

Orphans frequently visited his promiscuity and infidelity despite hate and love for tomorrow sex ain't nothing.

In the following examples, which were produced at the Bell Laboratories, the person adding a word saw three preceding words as well as the title:

About Life

Life has many good and wise men seldom condemn halfwits lightly! You wonder why not. Human feelings but savage tribes found . . .

Engineers

It is frequently said that they knew why forces might affect salaries. However, all scientists can't imagine . . .

Housecleaning

First empty the furniture of the master bedroom and bath. Toilets are to be washed after polishing doorknobs the rest of the room. Washing windows semiannually is to be taken by small aids such as husbands are prone to omit soap powder.

Murder Story

When I killed her I stabbed Claude between his powerful jaws clamped tightly together. Screaming loudly despite fatal consequences in the struggle for life ebbing as he coughed hollowly spitting blood from his ears.

I think that it is hard to read such material without amusement. I feel a little admiration as well. I would never write, "It happened one frosty look of trees waving gracefully against the wall." I almost wish I could. Poor poets endlessly rhyme love with dove, and they are constrained by their highly trained mediocrity *never* to produce a good line. In some sense, a stochastic process can do better; it at least has a chance. I wish I had hit on *deamy,* but I never would have.

Will a computer produce text of any literary merit by means of grammatical rules and a sequence of random numbers? It might produce fresh and amusing "words" and amusing short passages of some shock value. One can of course imagine a machine designed to write detective novels and equipped with settings for hard-boiled, puzzle, character, suspense, and so on, but such a device seems to me to be very far away.

The visual arts can be used to illustrate the same points which have been made in connection with music and language. A completely random visual pattern, like a completely random acoustic wave or a completely random sequence of letters, is mathematically the most surprising, the least predictable of all possible patterns. Alas, a completely random pattern is also the dullest of all patterns, and to a human being one random pattern looks just like another. Figure XIII-5, which is an array of 10,000 randomly black or white dots, illustrates this.

Bela Julesz, who works in the field of perception, caused an

Fig. XIII-5

electronic computer to produce this random array of dots as a part of his studies of stereoscopic vision and of the meaning of pattern. He also programmed the computer to remove some of the randomness from such a random pattern. He did this by making the computer examine successively various sets of five points located at the tips and at the center of an *X*, as shown by the points marked *X* in Figure XIII-6 (other points are marked O). If the center point was the same (black or white) as either points 1 and 4 or points 2 and 3, it was changed (from black to white or from white to black). This tends to remove any black or white diagonals, except when points 1 and 4 are black and points 2 and 3 are white or vice versa.

As we can see from Figure XIII-7, making a pattern less random in this way alters and improves its appearance profoundly. An unpredictable (random) component is desirable for the sake of variety or surprise, but some orderliness is necessary if a pattern is to be pleasing.

This exploitation of both order and randomness is in fact old to art. The kaleidoscope offers a charming effect by giving to a random arrangement of bits of colored glass a sixfold symmetry.

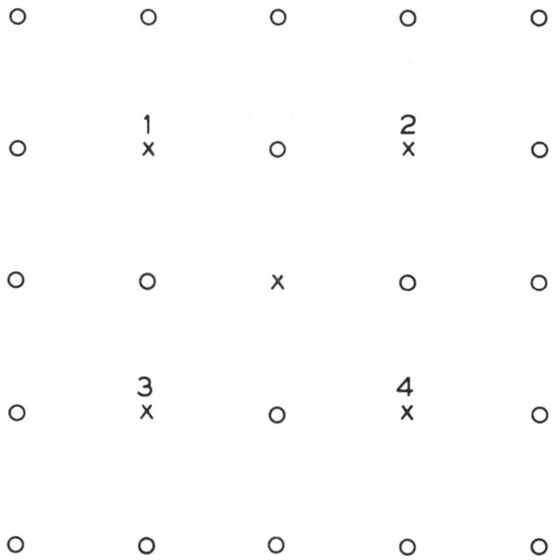

o	o	o	o	o
o	1 x	o	2 x	o
o	o	x	o	o
o	3 x	o	4 x	o
o	o	o	o	o

Fig. XIII-6

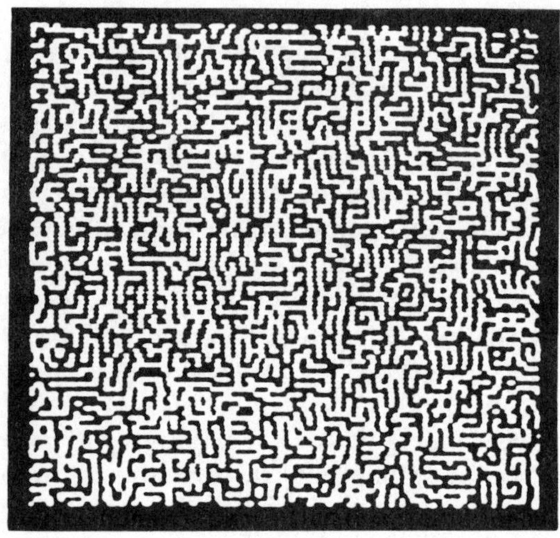

Fig. XIII-7

Many years ago Marcel Duchamp, who painted *Nude Descending a Staircase,* allowed a number of threads to fall on pieces of black cloth and then framed and preserved them. Jean Tinguley, the Swiss artist, has produced, by means of a machine, partly ordered, partly random colored designs of considerable merit; I derive continuing pleasure from one which hangs in my office. I saved for years a pile of solder droppings which I intended to mount on a block of ebony and present to the Museum of Modern Art. Finally, I lost both the solder and the desire to do so.

All of this has given me a sort of minimum philosophy of art, which I will not, I hasten to assure the reader, blame on information theory. It is a minimum philosophy because it says nothing about the talent or genius which alone can make art worth while.

Successful art requires the appreciation of an audience as well as the talent of the artist. Audiences are influenced by things other than the object of art before them. If a person sets his mind against it, anything will leave him cold. A desire to appreciate can, on the other hand, lead to one's liking even poor works. I like the hymn-like compositions that Betty Shannon and I made. Authors sometimes prefer inferior works of their own. Both small coteries and large groups can be led to appreciate sincerely things which are for a time the fashion but which have little long-range appeal and which probably have little merit.

Among other things, audiences want to have a sense of authorship, a sense of an individual, in connection with works of art. To bring appreciation to an artist, his work must have enough consistency so that it is recognizable as his. How let down the sincere appreciator must be if he always has to look at the label or wait for the announcer in order to know that the painting or music is the product of his favorite artist.

Suppose that one artist had actually produced in succession the masterpieces we now accept as the works of a number of great artists with diverse styles, long before the artists lived. This would astonish us, but we could scarcely appreciate him as an artist, however much we might admire the individual paintings. Picasso is eminently recognizable, but he is disquieting. He has been skillful in many styles, and yet he escapes our final judgment by going from one style to another. How much easier it is to appreciate Matisse.

To be appreciated by an audience, art must be intelligible to the audience. Even a good joke in Chinese will amuse few Americans, and certainly ten jokes in Chinese will be no more amusing than one. To a degree, to be appreciated art must be in a language familiar to the audience; otherwise no matter how great the variety may be, the audience will have an impression of monotony, of sameness. We can be surprised repeatedly only by constrast with that which is familiar, not by chaos.

Some artists adopt a language taught to their audience by earlier masters. Brahms was one of these. Other artists teach something of a new language to their audiences, as the impressionists did. Certainly, the language of art changes with time, and we should be grateful to the artists who teach us new words. However, we should not doubt the originality of such artists as Bach and Handel, who spoke ringingly in a language of the past.

While a language with intelligible words and relations between words is necessary in art, it is not sufficient. Mechanical sameness is dull and disappointing. I prefer the surprises of stochastic prose to the vapid verses of Owen Meredith. Perhaps in some age of bad art, man will be forced to stochastic art as an alternative to the stale product of human artisans.

So much for information theory and art.

CHAPTER XIV *Back to Communication Theory*

SURELY, IT IS WONDERFUL if a new idea contributes to the solution of a broad range of problems. But, first of all, to be worthy to notice a new idea must have some solid and clearly demonstrated value, however narrow that value may be.

An information theorist has criticized me for exploring in this book possible applications of information theory in fields of language, psychology, and art. To him, the relation between such subjects and information theory seems marginal or even dubious. Why distract the reader from the clearly demonstrated value and importance of information theory by discussing matters concerning which no clear value or importance can be demonstrated?

Partly, in writing this book I have felt an obligation to the reader to discuss relations between information theory in its solid and narrow sense and various fields with which it has been connected in the writings of others. Partly, I believe that information theory is useful in helping us in talking sense or at least in keeping from talking nonsense in connection with some linguistic, artistic, and psychological problems. However, there is a danger in overemphasizing such matters in a book on information theory.

It would certainly be wrong to assert or to believe that information theory is valuable chiefly because of wide-ranging connections with a variety of fields such as language, cybernetics, psychology, and art. To believe this would be to repeat mistakes which have been made in connection with other important discoveries.

Thus, in Newton's day his work was beclouded by controversy and philosophy, and for many years thereafter it was associated in people's minds with a putative universality which confused its real nature. Einstein, however, could see more clearly. He said: "Reason, of course, is weak when measured against its never ending task." Einstein then described Newton's contribution to this task of understanding and observed, "and with that, the goal was reached, the science of celestial mechanics was born, confirmed a thousand times over by Newton himself and those who came after him."

It is fair to add that since Newton's day, Newtonian mechanics has been useful in solving or contributing to the solution of problems that never entered the minds of Newton and his contemporaries, but it has not solved all problems of science, as some optimistic philosophers expected it to.

To me the indubitably valuable content of information theory seems clear and simple. It embraces the ideas of the information rate or entropy of an ergodic message source, the information capacity of noiseless and noisy channels, and the efficient encoding of messages produced by the source, so as to approach errorless transmission at a rate approaching the channel capacity. The world of which information theory gives us an understanding of clear and present value is that of electrical communication systems and, especially, that of intelligently designing such systems.

It seems to me wise at the close of this book to turn away from the broad, speculative possibilities (or impossibilities?) of information theory and to ask the following question: Beyond the things already described in this book, what have information theorists done and what are they doing that is mathematically sound, well founded, compelling? What, in other words, have they done that qualifies as sound science which we *must* accept rather than as intriguing speculation that we have the privilege of arguing about?

Here we find a broad range of work. To explain all of it fully to the reader would take another book. Thus, this chapter will be a brief summary of some of the work of information theorists since the publication of Shannon's original paper. Its purpose is to acquaint the reader with the scope of information theory in its narrow sense and, perhaps, to entice him into following such activities in greater detail.

One thing that information theorists have sought is some application of the entropy of information rate of a message source to a problem other than that of encoding and transmission of information. Ambitious men want to bring meaning into the picture somehow, but a more modest worker is willing to settle for any application which is meaningful and rigorously correct.

The only application of information rate to a problem other than efficient encoding which has been given so far and which meets these criteria was advanced by J. L. Kelly, Jr., in 1956.[1] It concerns gambling on chance events in which the bettor has inside information as to the outcome of the event bet upon. We might imagine, for instance, that the dice are already thrown (or the race run) and that the favored bettor knows this and has received some knowledge of the outcome, but the person with whom he bets doesn't know this and gives the bettor fair odds on the basis of the chance of the outcome.

The information which the bettor receives is doled out to him in bits, that is, yes-or-no answers to questions. His informant could, for instance, inform the bettor completely concerning whether a coin tossed had turned up heads or tails by sending him one bit of information. Or the informant could narrow for the bettor the possible outcomes of the cast of a die from 6 to 3 by using one bit of information to tell the bettor whether the outcome was odd or even.

Following this introduction, I can best explain Kelly's result by quoting the abstract of his paper:

> If the input symbols to a communication channel represent the outcomes of a chance event on which bets are available at odds consistent with their probabilities (i.e., "fair" odds), a gambler can use the knowledge given him by the received symbols to cause his money to grow exponentially. The maximum exponential rate of growth of the gambler's capital is equal to

[1] "New Interpretation of Information Rate," *Bell System Technical Journal,* Vol. 35 (July, 1956), pp. 917–926.

the rate of transmission of information over the channel. This result is generalized to include the case of arbitrary odds.

Thus we find a situation in which the transmission rate is significant even though no coding is contemplated. Previously this quantity was given significance only by a theorem of Shannon's which asserted that, with suitable encoding, binary digits could be transmiited over the channel at this rate with an arbitrarily small probability of error.

Numerically the factor by which the gambler's initial capital is increased after N bets is

$$2^{NR}$$

Here R is the average number of bits of information transmitted to the bettor per bet.

If this seems a trivial application of the amount of information in bits, the reader should meditate on the fact that it is the *only* mathematically established interpretation, other than those concerned with the rate of generation of probable messages and their efficient encoding for transmission, that anyone has discovered.

In advancing information theory, one may seek a new use for information theory rather than a new interpretation of information rate. Thus, in 1949, C. E. Shannon published a long paper entitled "Communication Theory of Secrecy Systems."[2] It is doubtful whether this paper has helped substantially in the deciphering of messages, but it has provided, for the first time, a well organized theory of cryptography and cryptanalysis, and it is highly regarded by the expert cryptanalysts.

It would be hopeless to try to go into the details of Shannon's work here, but I will try to give an idea of some of its content.

The cryptanalyst who lays hands on a message enciphered by an unknown means is ignorant of two things: the message itself and a specification of the means used to encipher it, which we may call the key.

Sometimes, the cryptanalyst may know the general scheme of encipherment. To take a ridiculously simple example, he might know that a simple substitution cipher had been used, that is, for each letter of the alphabet some other letter had been substituted according to a fixed scheme.

[2] *Ibid.*, Vol. 28 (October, 1949), pp. 656–715.

The cryptanalyst may have a short or a long enciphered message to work with. If the message had only three letters in it, say QXD, these might stand for AND, or BET, or any other English word made up of three different letters. As the message becomes longer, however, the number of possible English texts which could have been encrypted by means of a simple substitution cipher to give the particular message at hand decreases; if the enciphered message is long enough, there will be only one possible source message.

Shannon expressed this decrease of uncertainty as to what message might have been enciphered so as to give the message in question as a change in the equivocation. The equivocation $H_y(x)$ of Chapter VIII gives the uncertainty of what message was enciphered by the general means in question in order to give the received enciphered message. Shannon was able to compute in the case of various ciphers how the equivocation decreases as the number of characters in the message increases. When the equivocation approaches zero, only one message could have been enciphered to give the enciphered message, and, in principle, the message can be deciphered uniquely.

What other sorts of problems have confronted or now confront information theorists? Some of these problems concern the sampling theorem. Information theorists use the sampling theorem in order to represent a smoothly varying, band-limited signal by means of a sequence of numbers; the sample numbers are the amplitudes of the signal taken every $1/2W$ seconds, where W is the band width of the signal.

The samples which represent a given band-limited signal are not unique; they can be taken at various times. Thus, in Figure XIV-1, either the vertical solid lines or the vertical dashed lines are samples which legitimately represent the function, and samples could have been taken at many other locations. In fact, the samples don't even have to be equally spaced in time, provided that, on the average, there are two $2W$ samples per second!

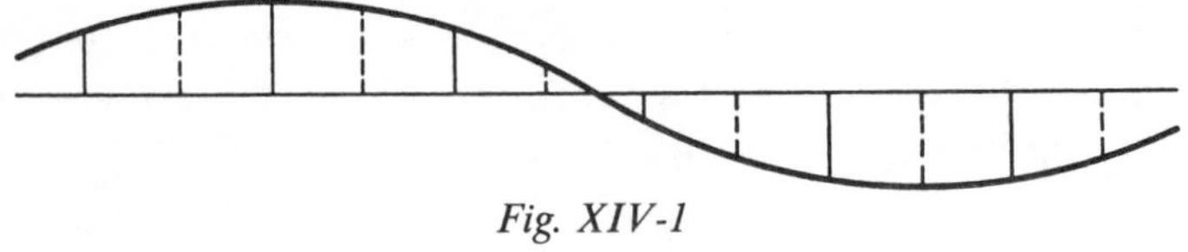

Fig. XIV-1

A band-limited signal is represented uniquely by $2W$ samples per second only when all samples from the infinite past to the infinite future are used. Sometimes we would like to talk about a piece of band-limited signal or about a band-limited signal which is almost zero except for some specified range of time, and we would like to describe such a portion of a signal or a signal of limited duration handily in terms of samples.

Our first thought might be, can we merely specify a short signal or a portion of a signal by specifying the values of a finite sequence of samples and saying nothing about samples before or after these? Alas, specifying such a finite set of samples does not specify just one band-limited signal; many different band-limited signals can be passed through a finite sequence of samples, and, if the signals are very large outside of the range of the specified samples, they can be very different within the range of the specified samples.

This failing, we might say, let us specify certain successive sample values and make all preceding and succeeding samples be zero. Surely, we may think, the band-limited signal so specified will conform closely to the sample values where these are not zero and will be small wherever the samples are specified as zero.

Suppose, for instance, that we insist that all of a set of equally spaced samples after a time t_0 are zero, while the samples before the time t_0 are nonzero, as shown by the dots in Figure XIV-2. Because the samples are specified for all times past and future, they do specify a unique band-limited signal. Will this signal be nearly zero for times after t_0?

Alas, H. O. Pollak, of the Bell Laboratories, has shown that this need not be so. Suppose we ask, what part of the total energy of

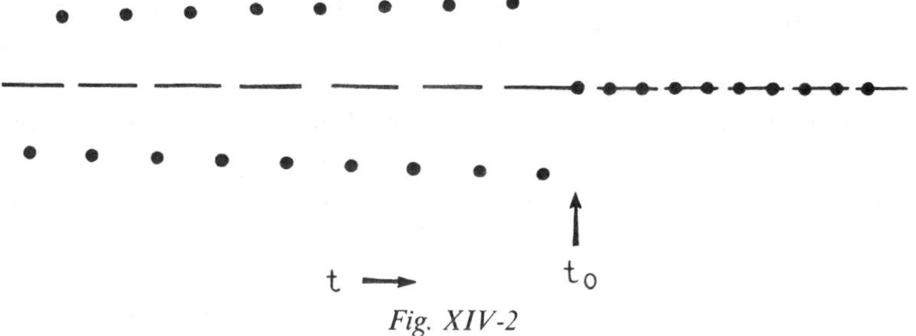

Fig. XIV-2

the band-limited signal passing through such samples is carried by the part of the wave which occurs ten seconds, or twenty minutes, or fifty years after t_0? Remember all the samples are zero after t_0.

The surprising answer is that almost half of the energy of the signal can be carried by the part that occurs later than *any specified time* after the samples become zero. Thus, the signal can be zero at all the samples after t_0 and still be large in between them.

Efforts to use the sampling theorem rigorously to represent signals of limited length are in mathematical trouble, and mathematicians are trying to find some way out.

Work by Pollak and Slepian indicates that neither samples nor sine waves are the most appropriate way to represent band-limited functions of finite duration, and these mathematicians have used a more appropriate group of functions called *prolate spheroidal functions* for this purpose.

One puzzling matter about information theory may be illustrated by the following example. Suppose that in telegraphy we let a positive pulse represent a dot and a negative pulse represent a dash. Suppose that some practical joker reverses connections so that when a positive pulse is transmitted a negative pulse is received and when a negative pulse is transmitted a positive pulse is received. Because no uncertainty has been introduced, information theory says that the rate of transmission of information is just the same as before. Yet we feel that some damage has been done to the communication system. The damage would be even more appalling if, in a teletypewriter link, we consistently printed out W for A, K for B, and so on, in a completely scrambled fashion.

This bothered Shannon, and he has worked out a theory to cover the situation. In this theory, he establishes a fidelity criterion. Thus, he might assign a given penalty for substituting a consonant for a vowel and a lesser penalty for substituting one vowel for another. He can then assess the damage done to a message by either consistent or random errors. When the damage is done by the random errors of a noisy channel, he shows in principle how to minimize it, and he shows how many bits per second are required to transmit the signal with a given degree of fidelity.

Shannon has also done a considerable amount of work concerning the transmission of messages over networks in which one message may interfere with another message. The simplest case is that of transmission of messages in both directions over the same channel between two points, *A* and *B*. As a very special case, we will assume that the circuit acts the same from *B* to *A* as from *A* to *B*.

Suppose that we plot the channel capacity for transmission from *A* to *B* against the channel capacity for transmission from *B* to *A*, as shown in Figure XIV-3. We can imagine two very simple cases. In one case, transmission from *B* to *A* does not interfere with transmission from *A* to *B*, and transmission from *A* to *B* does not interfere with transmission from *B* to *A*. In this case, the curve consists of the horizontal solid line giving the channel capactiy from *B* to *A* and the vertical solid line giving the channel capacity from *A* to *B*.

Or we can imagine that at one time we can transmit in one

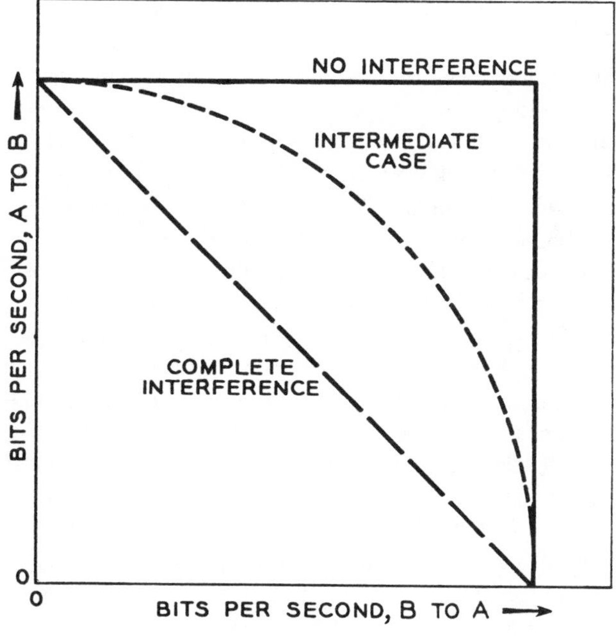

Fig. XIV-3

direction only, either from A to B or from B to A. Then if we are transmitting from A to B one-third of the time, we can transmit from B to A only two-thirds of the time, and so on. The sum of the channel capacity from B to A and the channel capacity from A to B must be a constant, and the result is the dashed 45° line of Figure XIV-3.

In an intermediate case, in which there is some interference between transmission in the two directions, we will get a curve roughly of the form of the dotted line of Figure XIV-3.

The study of efficient encoding continues to command the attention of information theorists. In the case of discrete channels, information theorists continually better codes for correcting A errors in a sequence of B digits.

Information theorists also seek best codes for transmitting information over a noisy continuous channel. In 1959, Shannon published a long paper in which he arrived at upper and lower bounds on the attainable error rates for codes of various complexity (that is, length) used in signaling over a continuous channel with Gaussian noise. Currently, convolutional codes and Viterbi decoding are favored, and ingenious men vie in making better and cheaper decoders.

Further, engineers who wish to improve electrical communication continually try to find new encoding and transmission schemes which are simple enough to be useful. They try to encode television and voice signals into as few binary digits per second as they can; the approaches they use have been indicated in Chapter VII. Such efficient encoding is growing in importance because digital transmission of signals (as in pulse code modulation) is displacing analog communication. It will grow in importance as the encrypting of signals in order to obtain privacy or secrecy becomes more common, for secrecy is best attained by digital means.

Engineers also look for simple and efficient error-correcting means useful in correcting the multiple errors which occur in the transmission of digital signals over existing telephone circuits. The use of digital transmission in transmitting text and in transmitting business and technical data is growing by leaps and bounds, both in military and in civilian applications. Telephone circuits go almost everywhere. To keep pace with data, we must use voice circuits for data. Here error correction by error detection and retransmission is the favored technique. But, error-correcting codes have their place.

There are special circumstances that call for special means of modulation. Mobile radio is one. In a city, signals reach the vehicle from many directions after bouncing off many buildings, and a short pulse would be received as a smear of pulses which have traveled different distances over different paths. Here great ingenuity is needed in finding the best way to use a large overall bandwidth in improving transmission.

Military communication, especially in the face of jamming, poses a host of problems.

Perhaps some would regard all this as engineering drudgery, unexciting compared with the broad philosophical vistas which information theory seems to open to us. Can an informed understanding, a loving appreciation of the nature, virtues and distinctions among the French Impressionists or the Dutch genre painters ever be so meaningful as a sudden and bewildering confrontation with a new and strange world of art, such as the Japanese?

Yet, the connoisseur who pursues with devotion the details of a field may well have as much insight and as sound values as the rapturous dilettante. There is some intellectual obligation to appreciate a field for what it is rather than for the reactions it excites in the minds of the uninformed. I hope that this book has its exciting aspects, but I also hope that it won't lead the reader to a view of information theory widely different from that held by informed workers in the field. Hence, it is perhaps well to end in a sober vein.

APPENDIX: *On Mathematical Notation*

THE READER WILL FIND a fairly liberal use of mathematical notation in this book, including a number of equations. This may incline him to say the book is full of mathematics.

Of course it is. Communication theory is a mathematical theory, and, as this book is an exposition of communication theory, it is bound to contain mathematics. The reader should not, however, confuse the mathematics with the notation used. The book could contain just as much mathematics and not include one symbol or equality sign.

The Babylonians and the Indians managed quite a lot of mathematics, including parts of algebra, without the aid of anything more than words and sentences. Mathematical notation came much later. Its purpose is to make mathematics easier, and it does for anyone who becomes familiar with it. It replaces long strings of words which would have to be used over and over again with simple signs. It provides convenient names for quantities that we talk about. It presents relations concisely and graphically to the eye, so that one can see at a glance the relations among quantities which would otherwise be strewn through sentences that the eye would be perplexed to comprehend as a whole.

The use of mathematical notation merely expresses or represents mathematics, just as letters represent words or notes represent music. Mathematical notation can represent nonsense or nothing, just as jumbled letters or jumbled notes can represent nothing. Crackpots often write tracts full of mathematical notation which stands for no mathematics at all.

In this book I have tried to put all the important ideas into words in sentences. But, because it is simpler and easier to understand things written concisely in mathematical notation, I have in most cases put statements into mathematical notation also. I have to a degree explained this throughout the book, but here I summarize and enlarge on these explanations. I have also ventured to include a few simple related matters which are not used elsewhere in this book, in the hope that these may be of some general use or interest to the reader.

The first thing to be noted is that letters can stand for numbers and for other things as well. Thus, in Chapter V, B_j stands for a group or sequence of symbols or characters, a group of letters perhaps; j signifies *which* group. For the first group of letters, j might be 1, and that first group might be AAA, for instance. For another value of j, say, 121, the group of letters might be ZQE.

We often have occasion to add, subtract, multiply, or divide numbers. Sometimes we represent the numbers by letters. Examples of the notations for these operations are:

Addition

$$2 + 3$$
$$a + d$$

We read $a + d$ as "a plus d." We may interpret $a + d$ as the sum of the number represented by a and the number represented by d.

Subtraction

$$5 - 4$$
$$q - r$$

We read $q - r$ as "q minus r."

Multiplication

$$3 \times 5 \text{ or } 3 \cdot 5 \text{ or } (3)(5)$$
$$u \times v \text{ or } u \cdot v \text{ or } uv$$

If we did not use parentheses to separate 3 and 5 in (3)(5), we would interpret the two digits as 35 (thirty-five). We can use parentheses to distinguish any quantities we want to multiply. We could write uv as $(u)(v)$, but we don't need to. We read (3) (5) as 3 times 5, but we read uv as "uv" with no pause between the u and v, rather than as "u times v."

Division

$$6 \div 3 \text{ or } \frac{6}{3} \text{ or } 6/3$$

$$\frac{1}{p} \text{ or } 1/p$$

We ordinarily read $1/p$ as "1 over p" rather than as "1 divided by p."

Quantities included in parentheses are treated as one number; thus

$$\frac{(2+4)}{3} = \frac{6}{3} = 2$$
$$(4+8)(2) = (12)(2) = 24$$
$$(a+b)c = ac + bc$$

We read $(a + b)$ either as "a plus b" or as "the quantity a plus b," if just saying "a plus b" might lead to confusion. Thus, if we said "c times a plus b" we might mean $ca + b$, though we would read $ca + b$ as "ca plus b." If we say "c times the quantity a plus b," it is clear that we mean $c(a + b)$.

The idea of a probability is used frequently in this book. We might say, for instance, that in a string of symbols the probability of the jth symbol is $p(j)$. We read this "p of j."

The symbols might be words, numbers, or letters. We can imagine that the symbols are tabulated; various values of j can be taken as various numbers which refer to the symbols. Table XVI, shows one way in which the numbers j can be assigned to the letters of the alphabet.

When we wish to refer to the probability of a particular letter, *N* for instance, we could, I suppose, refer to this as $p(5)$, since 5 refers to N in the above table. We'd ordinarily simply write p(N), however.

What is this probability? It is the fraction of the number of

TABLE XVI

Value of j	*Corresponding Letter*
1	E
2	T
3	A
4	O
5	N
6	R
etc.	

letters in a long passage which are the letter in question. Thus, out of a million letters, close to 130,000 will be E's, so

$$p(\mathrm{E}) = \frac{130{,}000}{1{,}000{,}000} = .13$$

Sometimes we speak of probabilities of two things occurring together, either in sequence or simultaneously. For instance, x may stand for the letter we send and y for the letter we receive. $p(x, y)$ is the probability of sending x and receiving y. We read this "*p* of x, y (we represent the comma by a pause). For instance, we might send the particular letter W and receive the particular letter B. The probability of this particular event would be written $p(\mathrm{W, B})$. Other particular examples of $p(x, y)$ are $p(\mathrm{A, A})$, $p(\mathrm{Q, S})$, $p(\mathrm{E, E})$, etc. $p(x, y)$ stands for all such instances.

We also have *conditional* probabilities. For instance, if I transmit x, what is the probability of receiving y? We write this conditional probability $p_x(y)$. We read this "*p* sub *x* of *y*." Many authors write such a conditional probability $p(y \mid x)$, which can be read as, "the probability of y given x." I have used the same notation which Shannon used in his original paper on communication theory.

Let us now write down a simple mathematical relation and interpret it:

$$p(x, y) = p(x)\, p_x(y)$$

That is, the probability of encountering x and y together is the probability of encountering x times the probability of encountering y, when we do encounter x. Or it may seem clearer to say that the number of times we find x and y together must be the number of times we find x times the fraction of times that y, rather than some other letter, is associated with x.

We frequently want to add many things up; we represent this by means of the summation sign Σ, which is the Greek letter sigma. Suppose that j stands for an integer, so that j may be 0, 1, 2, 3, 4, 5, etc. Suppose we want to represent

$$0 + 1 + 2 + 3 + 4 + 5 + 6 + 7 + 8$$

which of course is equal to 36. We write this

$$\sum_{j=0}^{j=8} j$$

We read this, "the sum of j from j equals 0 to j equals 8." The Σ sign means sum. The $j = 0$ at the bottom means to start with 0, and the $j = 8$ at the top means to stop with 8. The j to the right of the sign means that what we are summing is just the integers themselves.

We might have a number of quantities for which j merely acts as a label. These might be the probabilities of various letters, for instance, according to Table XVII.

If we wanted to sum these probabilities for all letters of the alphabet we would write

$$\sum_{j=1}^{26} p(j)$$

We read this "the sum of $p(j)$ from j equals 1 to 26." This quantity is of course equal to 1. The fraction of times A occurs per letter plus the fraction of times B occurs per letter, and so on, is the fraction of times per letter that any letter at all occurs, and one letter occurs per letter.

If we just write

$$\sum_{j} p(j)$$

TABLE XVII

Value of j	*Letter Referred to*	*Probability of Letter, p (j)*
1	E	.13105
2	T	.10468
3	A	.08151
4	O	.07995
5	N	.07098
6	R	.06882
7	I	.06345
8	S	.06101
9	H	.05259
10	D	.03788
11	L	.03389
12	F	.02924
13	C	.02758
14	M	.02536
15	U	.02459
16	G	.01994
17	Y	.01982
18	P	.01982
19	W	.01539
20	B	.01440
21	V	.00919
22	K	.00420
23	X	.00166
24	J	.00132
25	Q	.00121
26	Z	.00077

it means to sum for all values of j, that is, for all that represent something. We read this, "the sum of $p(j)$ over j." If j is a letter of the alphabet, then we will sum over, that is, add up, twenty-six different probabilities.

Sometimes we have an expression involving two letters, such as i and j. We may want to sum with respect to one of these *indices*. For instance, $p(i, j)$ might be the probability of letter i occurring followed by letter j, as, $p(\text{Q}, \text{V})$ would be the probability of encountering the sequence QV. We could write, for instance

$$\sum_j p(i, j)$$

We read this, "the sum of p of i, j with respect to (or, over) j." This says, let j assume every possible value and add the probabilities. We note that

$$\sum_j p(i, j) = p(i)$$

This reads, "the sum of p of i, j over j equals p of i." If we add up the probabilities of a letter followed by every possible letter we get just the probability of the letter, since every time the letter occurs it is followed by *some* letter.

Besides addition, subtraction, multiplication, and division we also want to represent a number or quantity multiplied by itself some number of times. We do this by writing the number of times the quantity is to be multiplied by itself above and to the right of the quantity; this number is called an *exponent.*

$$2^1 = 2$$

"2 to the first (or 2 to the first power) equals 2." 1 is the exponent.

$$2^2 = 4$$

"2 squared, (or 2 to the second) equals 4." 2 is the exponent.

$$2^3 = 8$$

"2 cubed (or 2 to the third) equals 8." 3 is the exponent.

$$2^4 = 16$$

"2 to the fourth equals sixteen." 4 is the exponent.

We can let the exponent be a letter, n; thus, 2^n, which we read "2 to the n," means multiply 2 by itself n times. a^n, which we read "a to the n," means multiply a by itself n times.

To get consistent mathematical results we must say

$$a^0 = 1$$

"a to the zero equals 1," regardless of what number a may be.

Mathematics also allows fractional and negative exponents. We should particularly note that

$$a^{-n} = \frac{1}{a^n} \text{ or } 1/a^n$$

We read a^{-n} as "a to the minus n." We read $1/a^n$ as "one over a to the n."

It is also worth noting that

$$a^n a^m = a^{(n+m)}$$

"a to the n, a to the m equals a to the n plus m." Thus

$$2^3 \times 2^2 = 8 \times 4 = 32 = 2^5$$

or

$$4^{1/2} \times 4^{1/2} = 4^1 = 4.$$

A quantity raised to the ½ power is the *square root*

$$4^{1/2} = \text{the square root of } 4 = 2.$$

It is convenient to represent large numbers by means of the powers of 10 or some other number

$$3.5 \times 10^6 = 3{,}500{,}000$$

This is read "three point five times ten to the sixth, (or ten to the six)."

The only other mathematical function which is referred to extensively in this book is the logarithm. Logarithms can have different *bases.* Except in instances specifically noted in Chapter X, all the logarithms in this book have the base 2. The logarithm to the base 2 of a number is the power to which 2 must be raised to equal the number. The logarithm of any number x is written log x and read "log x." Thus, the definition of the logarithm to the base 2, as given above, is expressed mathematically by:

$$2^{\log x} = x$$

That is, "2 to the log x equals x."

As an example

$$\log 8 = 3$$
$$2^3 = 8$$

Other logarithms to the base 2 are

x	$\log x$
1	0
2	1
4	2
8	3
16	4
32	5
64	6

Some important properties of logarithms should be noted:

$$\log ab = \log a + \log b$$
$$\log a/b = \log a - \log b$$
$$\log d^c = c \log d$$

As a special case of the last relation,

$$\log 2^m = m \log 2 = m$$

Except in information theory, logarithms to the base 2 are not used. More commonly, logarithms to the base 10 or the base e ($e = 2.718$ approximately) are used.

Let us for the moment write the logarithm of x to the base 2 as $\log_2 x$, the logarithm to the base 10 as $\log_{10} x$, and the logarithm to the base e as $\log_e x$. It is useful to note that

$$\log_2 x = (\log_2 10)(\log_{10} x) = \frac{\log_{10} x}{\log_{10} 2}$$
$$\log_2 x = 3.32 \log_{10} x$$
$$\log_2 x = (\log_2 e)(\log_e x) = \frac{\log_e x}{\log_e 2}$$
$$\log_2 x = 1.44 \log_e x$$

The logarithm to the base e is called the *natural* logarithm. It has a number of simple and important mathematical properties. For instance, if x is much smaller than 1, then approximately

$$\log_e(1 + x) = x$$

Use is made of this approximation in Chapter X.

In the text of the book, by $\log x$ we always mean $\log_2 x$.

Glossary

ADDRESS: In a computer, a number designating a part of the memory used to store a number, also the part of the memory which is used to store a number.

ALPHABET: The alphabet, the alphabet plus the space, any given set of symbols or signals from which messages are constructed.

AMPLITUDE: Magnitude, intensity, height. The amplitude of a sine wave is its greatest departure from zero, its greatest height above or below zero.

ATTENUATION: Decrease in the amplitude of a sine wave during transmission.

AUTOMATON: A complicated and ingenious machine. Elaborate clocks which parade figures on the hour, automatic telephone switching systems, and electronic computers are automata.

AXIS: One of a number of mutually perpendicular lines which constitute a coordinate system.

BAND: A range or strip of frequencies.

BAND LIMITED: Having no frequencies lying outside of a certain band of frequencies.

BAND WIDTH: The width of a band of frequencies, measured in cps.

BINARY DIGIT: A 0 or a 1. 0 and 1 are the binary digits.

BIT: The choice between two equally probable possibilities.

BLOCK: A sequence of symbols, such as letters or digits.

BLOCK ENCODING: Encoding a message for transmission, not letter by letter or digit by digit, but, rather, encoding a sequence of symbols together.

BOLTZMANN'S CONSTANT: A constant important in radiation and other thermal phenomena. Boltzmann's constant is designated by the letter k. $k = 1.37 \times 10^{-23}$ joules per degree Celsius (centigrade).

BROWNIAN MOTION: Erratic motion of very small particles caused by the impacts of the molecules of a liquid or gas.

CAPACITOR: An electrical device or circuit element which is made up of two metal sheets, usually of thin metal foil, separated by a thin dielectric (insulating) layer. A capacitor stores electric charge.

CAPACITY: The capacity of a communication channel is equal to the number of bits per second which can be transmitted by means of the channel.

CHANNEL VOCODER: A vocoder in which the speech is analyzed by measuring its energy in a number of fixed frequency ranges or bands.

CHECK DIGITS: Symbols sent in addition to the number of symbols in the original message, in order to make it possible to detect the presence of or correct errors in transmission.

CLASSICAL: Prequantum or prerelativistic.

COMMAND: One of a number of elementary operations a computer can carry out, e.g., add, multiply, print out, and so on.

COMPLICATED MACHINE: An automaton.

CONTACT: A piece of metal which can be brought into contact with another piece of metal (another contact) in order to close an electric circuit.

COORDINATE: A distance of a point in a space from the origin in a direction parallel to an axis. In three-dimensional space, how far up or down, east or west, north or south a point is from a specified origin.

CORE (magnetic): A closed loop of magnetic material linked by wires. Cores are used in the memory of an electronic computer. Magnetization one way around the core means 1; magnetization the other way around the core means 0.

CPS: Cycles per second, the terms in which frequency is measured.

CYCLE: A complete variation of a sine wave, from maximum, to minimum, to maximum again.

DELAY: The difference between the time a signal is received and the time it was sent.

DETECTION THEORY: Theory concerning when the presence of a signal can be determined even though the signal is mixed with a specified amount of noise.

DIGRAM PROBABILITY: The probability that a particular letter will follow another particular letter.

DIMENSION: The number of numbers or coordinates necessary to specify the position in a space is the number of dimensions in the space. The space of experience has three dimensions: up–down, east–west, north–south.

DIODE: A device which will conduct electricity in one direction but not in the other direction.

DISCRETE SOURCE: A message source which produces a sequence of symbols such as letters or digits, rather than an electric signal which may have any value at a given time.

DISTORTIONLESS: Transmission is distortionless if the attenuation is the same for sine waves of all frequencies and if the delay is the same for sine waves of all frequencies.

DOUBLE-CURRENT TELEGRAPHY: Telegraphy in which use is made of three distinct conditions: no current, current flowing into the wire, and current flowing out of the wire.

ELECTROMAGNETIC WAVE: A wave made up of changing electric and magnetic fields. Light and radio waves are electromagnetic waves.

ENERGY LEVEL: According to quantum mechanics, a particle (atom, electron) cannot have *any* energy, but only one of many particular energies. A particle is in a particular energy level when it has the energy and motion characteristic of that energy level.

ENSEMBLE: All of an infinite number of things taken together, such as, all the messages that a given message source can produce.

ENTROPY: The entropy of communication theory, measured in bits per symbol or bits per second, is equal to the average number of binary digits per symbol or per second which are needed in order to transmit messages produced by the source. In communication theory, entropy is interpreted as average uncertainty or choice, e.g., the average uncertainty as to what symbol the source will produce next or the average choice the source has as to what symbol it will produce next. The entropy of statistical mechanics measures the uncertainty as to which of many possible states a physical system is actually in.

EQUIVOCATION: The uncertainty as to what symbols were transmitted when the received symbols are known.

ERGODIC: A source of text is ergodic if each ensemble average, taken over all messages the source can produce, is the same as the corresponding average taken over the length of a message. *See* Chapter III.

FILTER: An electrical network which attenuates sinusoidal signals of some frequencies more than it attenuates sinusoidal signals of other frequencies. A filter may transmit one band of frequencies and reject all other frequencies.

FINITE-STATE MACHINE: A machine which has only a finite number of different states or conditions. A switch which can be set at any of ten positions is a very simple finite-state machine. A pointer which can be set at any of an infinite number of positions is *not* a finite-state machine.

FM: Frequency modulation, representing the amplitude of a signal to be transmitted by the frequency of the wave which is transmitted.

FORMANT: In speech sounds, there is much energy in a few ranges of frequency. Strong energy in a particular range of frequencies in a speech sound constitutes a formant. There are two or three principal formants in speech.

FREQUENCY: The reciprocal of the period of a sine wave; the number of peaks per second.

GALVANOMETER: A device used to detect or measure weak electric currents.

GAUSSIAN NOISE: Noise in which the chance that the intensity measured at any time has a certain value follows one very particular law.

HYPERCUBE: The multidimensional analog of a cube.

HYPERSPHERE: The multidimensional analog of a sphere.

INDUCTOR: An electric device or circuit element made up of a coil of highly conducting wire, usually copper. The coil may be wound on a magnetic core. An inductor resists changes in electric current.

INPUT SIGNAL: The signal fed into a transmission system or other device.

JOHNSON NOISE: Electromagnetic noise emitted from hot bodies; thermal noise.

JOULE: A measure or amount of energy or work.

LATENCY: Interval of time between a stimulus and the response to it.

LINE SPEED: The rate at which distinct, different current values can be transmitted over a telegraph circuit.

LINEAR: An electric circuit or any system or device is linear if the response to the sum of two signals is the sum of the responses which would have been obtained had the signals been applied separately. If the output of a device at a given time can be expressed as the sum of products of inputs at previous times and constants which depend only on remoteness in time, the device is necessarily linear.

LINEAR PREDICTION: Prediction of the future value of a signal by means of a linear device.

MAP: To assign on one diagram a point corresponding to every point on another diagram.

MAXWELL'S DEMON: A hypothetical and impossible creature who, without expenditure of energy, can see a molecule coming in a gas which is all at one temperature and act on the basis of this information.

MEMORY: The part of an electronic computer which stores or remembers numbers.

MESSAGE: A string of symbols; an electric signal.

MESSAGE SOURCE: A device or person which generates messages.

NEGATIVE FEEDBACK: The use of the output of a device to change the input in such a way as to reduce the difference between the input and a prescribed input.

NEGATIVE FEEDBACK AMPLIFIER: An amplifier in which negative feedback is used in order to make the output very nearly a constant times the input, despite imperfections in the tubes or transistors used in the amplifier.

NETWORK: An interconnection of resistors, capacitors, and inductors.

NOISE: Any undesired disturbance in a signaling system, such as, random electric currents in a telephone system. Noise is observed as static or hissing in radio receivers and as "snow" in TV.

NOISE TEMPERATURE: The temperature a body would have to have in order to emit Johnson noise of any intensity equal to the intensity of an observed or computed noise.

NONLINEAR PREDICTION: Prediction of the future value of a signal by means of a nonlinear device, that is, any device which is not linear.

ORIGIN: The point at which the axes of a coordinate system intersect.

OUTPUT SIGNAL: The signal which comes out of a transmission system or device.

PERIOD: The time interval between two successive peaks of a sine wave.

PERIODIC: Repeating exactly and regularly time after time.

PERPETUAL MOTION: Obtaining limitless mechanical energy or work contrary to physical laws. Perpetual-motion machines of the first kind would generate energy without source. Perpetual-motion machines of the second kind would turn the unavailable energy of the heat of a body which is all at one temperature into ordered mechanical work or energy.

PHASE: A measure of the time at which a sine wave reaches its greatest height. The phase *angle* between two sine waves of the same frequency is proportional to the fraction of the period separating their peak values.

PHASE SHIFT: Delay measured as a fraction of the period rather than as a time difference.

PHASE SPACE: A multidimensional space in which the velocity and the position of each particle of a physical system is represented by distance parallel to a separate axis.

PHONEME: A class of allied speech sounds, the substitution of one of which for another in a word will not cause a change in meaning. The sounds of b and p are different phonemes, the substitution of one of which for another can change the meaning of a word.

POTENTIAL THEORY: The mathematical study of certain equations and their solutions. The results apply to gravitational fields, to certain aspects of electric and magnetic fields, and to certain aspects of the flow of air and liquids.

POWER: Rate of doing work or of expending energy. A watt is 1 joule per second.

PROBABILITY: In mathematics, a number between 0 and 1 associated with an event. In applications this number is the fraction of times the event occurs in many independent repetitions of an experiment. E.g., the probability that an ideal, unbiased coin will turn up heads is .5.

QUANTUM: A small, discrete amount of energy and, especially, of electromagnetic energy.

QUANTUM THEORY: Physical theory that takes into account the fact that energy and other physical quantities are observed in discrete amounts.

RADIATE: To emit electromagnetic waves.

RADIATION: Electromagnetic waves emitted from a hot body (anything above absolute zero temperature).

RANDOM: Unpredictable.

REDUNDANT: A redundant signal contains detail not necessary to determine the intent of the sender. If each digit of a number is sent twice (1 1 0 0 1 1 instead of 1 0 1) the signal or message is redundant.

REGISTER: In a computer, a special memory unit into which numbers to be operated on (to be added, for instance) are transferred.

RELAY: An electrical device consisting of an electromagnet, a magnetic bar which moves when the electromagnet is energized, and pairs of contacts which open or close when the bar moves.

RESISTOR: An electrical device or circuit element which may be a coil of fine poorly conducting wire, a thin film of poorly conducting material, such as carbon, or a rod of poorly conducting material. A resistor resists the flow of electric current.

SAMPLE: The value or magnitude of a continuously varying signal at a particular specified time.

SAMPLING THEOREM: A signal of band width W cps is perfectly specified or described by its exact values at $2W$ equally spaced times per second.

SERVOMECHANISM: A device which acts on the basis of information received to change the information which will be received in the future in accordance with a specific goal. A thermostat which measures the temperature of a room and controls the furnace to keep the temperature at a given value is a servomechanism.

SIGN: In medicine, something which a physician can observe, such as an elevated temperature. In linguistics, a pictograph or other imitative drawing.

SIGNAL: Any varying electric current deliberately transmitted by an electrical communication system.

SINE WAVE: A smooth, never-ending rising and falling mathematical curve. A plot vs. time of the height of a crank attached to a shaft which rotates at a constant speed is a sine wave.

SINGLE-CURRENT TELEGRAPHY: Telegraphy in which use is made of two distinct conditions: no current and current flowing into or out of the wire.

SPACE: A real or imaginary region in which the position of an object can be specified by means of some number of coordinates.

STATIONARY: A machine, or process, or source of text is stationary, roughly, if its properties do not change with time. *See* Chapter III.

STATISTICAL MECHANICS: Provides an explanation of the laws of thermodynamics in terms of the average motions of many particles or the average vibrations of a solid.

STATISTICS: In mathematical theories, we can specify or assign probabilities to various events. In judging the bias of an actual coin, we collect data as to how many times heads and tails turn up, and on the basis of these data we make a somewhat imperfect statistical estimate of the probability that heads will turn up. Statistics are estimates of probability on the basis of data. More loosely, "the statistics of a message source" refers to all the probabilities which describe or characterize the source.

STOCHASTIC: A machine or any process which has an output, such as letters or numbers, is stochastic if the output is in part dependent on truly random or unpredictable events.

STORE: Memory.

SUBJECT: A human animal on which psychological experiments are carried out.

SYMBOL: A letter, digit, or one of a group of agreed upon marks. Linguists distinguish a symbol, whose association with meaning or objects is arbitrary, from a *sign*, such as a pictograph of a waterfall.

SYMPTOM: In medicine, something that the physician can know only through the patient's testimony, such as, a headache, as opposed to a sign.

SYSTEM: In engineering, a collection of components or devices intended to perform some over-all functions, such as, a telephone switching system. In thermodynamics and statistical mechanics, a particular collection of material bodies and radiation which is under consideration, such as, the gas in a container.

TESSARACT: The four-dimensional analog of a cube, a hypercube of four dimensions.

THEOREM: A statement whose truth has been demonstrated by an argument based on definitions and on assumptions which are taken to be true.

THERMAL NOISE: Johnson noise.

THERMODYNAMICS: The branch of science dealing with the transformation of heat into mechanical work and related matters.

TOTAL ENERGY: The total energy of a signal is its average power times its duration.

TRANSISTOR: An electronic device making use of electron flow in a solid, which can amplify signals and perform other functions.

VACUUM TUBE: An electronic device making use of electron flow in a vacuum, which can amplify signals and perform other functions.

VOCODER: A speech transmission system in which a machine at the transmitting end produces a description of the speech; the speech itself is not transmitted, but the description is transmitted, and the description is used to control an artificial speaking machine at the receiving end which imitates the original speech.

WATT: A power of 1 joule per second.

WAVEGUIDE: A metal tube used to transmit and guide very short electromagnetic waves.

WHITE NOISE: Noise in which all frequencies in a given band have equal powers.

ZIPF'S LAW: An empirical rule that the number of occurrences of a word in a long stretch of text is the reciprocal of the order of frequency of occurrence. For example, the hundredth most frequent word occurs approximately 1/100 as many times as the most frequent word.

Index

About the Author

Dr. John R. Pierce was born in Des Moines, Iowa, in 1910 and spent his early life in the Midwest. He received his undergraduate education at the California Institute of Technology in Pasadena, and his M.A. and Ph.D. degrees in electrical engineering from the same institution.

In 1936 he went to the Bell Telephone Laboratories, and was Executive Director, Communication Sciences Division when he left in 1971 to become Professor of Engineering at the California Institute of Technology. He is now Chief Technologist at Caltech's Jet Propulsion Laboratory.

Dr. Pierce's writings have appeared in *Scientific American, The Atlantic Monthly, Coronet,* and several science fiction magazines. His books include *Man's World of Sound* (with E. E. David); *Electrons' Waves and Messages; Waves and the Ear* (with W. A. van Bergeijk and E. E. David); and several technical books, including *Traveling Wave Tubes, Theory and Design of Electron Beams, Almost All about Waves* and *An Introduction to Communication Science and Systems* (with E. C. Posner).

Dr. Pierce is a member of several scientific societies, including the National Academy of Sciences, the National Academy of Engineering, the American Academy of Arts and Sciences, the Royal Academy of Sciences (Sweden), and is a Fellow of The Institute of Electrical and Electronic Engineers, the American Physical Society and the Acoustical Society of America. He has received ten honorary degrees and a number of awards, including the National Medal of Science, the Edison Medal and the Medal of Honor (IEEE), the Founders Award (National Academy of Engineering), the Cedegren Medal (Sweden) and the Valdemar Poulson Medal (Denmark).

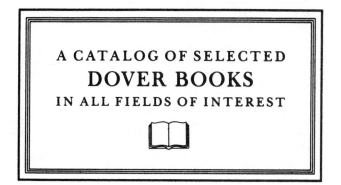

A CATALOG OF SELECTED

DOVER BOOKS

IN ALL FIELDS OF INTEREST

A CATALOG OF SELECTED DOVER BOOKS IN ALL FIELDS OF INTEREST

STICKLEY CRAFTSMAN FURNITURE CATALOGS, Gustav Stickley and L. & J. G. Stickley. Beautiful, functional furniture in two authentic catalogs from 1910. 594 illustrations, including 277 photos, show settles, rockers, armchairs, reclining chairs, bookcases, desks, tables. 183pp. 6½ x 9¼. 23838-5

AMERICAN LOCOMOTIVES IN HISTORIC PHOTOGRAPHS: 1858 to 1949, Ron Ziel (ed.). A rare collection of 126 meticulously detailed official photographs, called "builder portraits," of American locomotives that majestically chronicle the rise of steam locomotive power in America. Introduction. Detailed captions. xi+ 129pp. 9 x 12. 27393-8

AMERICA'S LIGHTHOUSES: An Illustrated History, Francis Ross Holland, Jr. Delightfully written, profusely illustrated fact-filled survey of over 200 American lighthouses since 1716. History, anecdotes, technological advances, more. 240pp. 8 x 10¾. 25576-X

TOWARDS A NEW ARCHITECTURE, Le Corbusier. Pioneering manifesto by founder of "International School." Technical and aesthetic theories, views of industry, economics, relation of form to function, "mass-production split" and much more. Profusely illustrated. 320pp. 6⅛ x 9¼. (Available in U.S. only.) 25023-7

HOW THE OTHER HALF LIVES, Jacob Riis. Famous journalistic record, exposing poverty and degradation of New York slums around 1900, by major social reformer. 100 striking and influential photographs. 233pp. 10 x 7⅞. 22012-5

FRUIT KEY AND TWIG KEY TO TREES AND SHRUBS, William M. Harlow. One of the handiest and most widely used identification aids. Fruit key covers 120 deciduous and evergreen species; twig key 160 deciduous species. Easily used. Over 300 photographs. 126pp. 5⅜ x 8½. 20511-8

COMMON BIRD SONGS, Dr. Donald J. Borror. Songs of 60 most common U.S. birds: robins, sparrows, cardinals, bluejays, finches, more–arranged in order of increasing complexity. Up to 9 variations of songs of each species.
Cassette and manual 99911-4

ORCHIDS AS HOUSE PLANTS, Rebecca Tyson Northen. Grow cattleyas and many other kinds of orchids–in a window, in a case, or under artificial light. 63 illustrations. 148pp. 5⅜ x 8½. 23261-1

MONSTER MAZES, Dave Phillips. Masterful mazes at four levels of difficulty. Avoid deadly perils and evil creatures to find magical treasures. Solutions for all 32 exciting illustrated puzzles. 48pp. 8¼ x 11. 26005-4

MOZART'S DON GIOVANNI (DOVER OPERA LIBRETTO SERIES), Wolfgang Amadeus Mozart. Introduced and translated by Ellen H. Bleiler. Standard Italian libretto, with complete English translation. Convenient and thoroughly portable–an ideal companion for reading along with a recording or the performance itself. Introduction. List of characters. Plot summary. 121pp. 5¼ x 8½. 24944-1

TECHNICAL MANUAL AND DICTIONARY OF CLASSICAL BALLET, Gail Grant. Defines, explains, comments on steps, movements, poses and concepts. 15-page pictorial section. Basic book for student, viewer. 127pp. 5⅜ x 8½. 21843-0

THE CLARINET AND CLARINET PLAYING, David Pino. Lively, comprehensive work features suggestions about technique, musicianship, and musical interpretation, as well as guidelines for teaching, making your own reeds, and preparing for public performance. Includes an intriguing look at clarinet history. "A godsend," *The Clarinet,* Journal of the International Clarinet Society. Appendixes. 7 illus. 320pp. 5⅜ x 8½. 40270-3

HOLLYWOOD GLAMOR PORTRAITS, John Kobal (ed.). 145 photos from 1926-49. Harlow, Gable, Bogart, Bacall; 94 stars in all. Full background on photographers, technical aspects. 160pp. 8⅜ x 11¼. 23352-9

THE ANNOTATED CASEY AT THE BAT: A Collection of Ballads about the Mighty Casey/Third, Revised Edition, Martin Gardner (ed.). Amusing sequels and parodies of one of America's best-loved poems: Casey's Revenge, Why Casey Whiffed, Casey's Sister at the Bat, others. 256pp. 5⅜ x 8½. 28598-7

THE RAVEN AND OTHER FAVORITE POEMS, Edgar Allan Poe. Over 40 of the author's most memorable poems: "The Bells," "Ulalume," "Israfel," "To Helen," "The Conqueror Worm," "Eldorado," "Annabel Lee," many more. Alphabetic lists of titles and first lines. 64pp. 5$\frac{3}{16}$ x 8¼. 26685-0

PERSONAL MEMOIRS OF U. S. GRANT, Ulysses Simpson Grant. Intelligent, deeply moving firsthand account of Civil War campaigns, considered by many the finest military memoirs ever written. Includes letters, historic photographs, maps and more. 528pp. 6⅛ x 9¼. 28587-1

ANCIENT EGYPTIAN MATERIALS AND INDUSTRIES, A. Lucas and J. Harris. Fascinating, comprehensive, thoroughly documented text describes this ancient civilization's vast resources and the processes that incorporated them in daily life, including the use of animal products, building materials, cosmetics, perfumes and incense, fibers, glazed ware, glass and its manufacture, materials used in the mummification process, and much more. 544pp. 6⅛ x 9¼. (Available in U.S. only.) 40446-3

RUSSIAN STORIES/RUSSKIE RASSKAZY: A Dual-Language Book, edited by Gleb Struve. Twelve tales by such masters as Chekhov, Tolstoy, Dostoevsky, Pushkin, others. Excellent word-for-word English translations on facing pages, plus teaching and study aids, Russian/English vocabulary, biographical/critical introductions, more. 416pp. 5⅜ x 8½. 26244-8

PHILADELPHIA THEN AND NOW: 60 Sites Photographed in the Past and Present, Kenneth Finkel and Susan Oyama. Rare photographs of City Hall, Logan Square, Independence Hall, Betsy Ross House, other landmarks juxtaposed with contemporary views. Captures changing face of historic city. Introduction. Captions. 128pp. 8¼ x 11. 25790-8

AIA ARCHITECTURAL GUIDE TO NASSAU AND SUFFOLK COUNTIES, LONG ISLAND, The American Institute of Architects, Long Island Chapter, and the Society for the Preservation of Long Island Antiquities. Comprehensive, well-researched and generously illustrated volume brings to life over three centuries of Long Island's great architectural heritage. More than 240 photographs with authoritative, extensively detailed captions. 176pp. 8¼ x 11. 26946-9

NORTH AMERICAN INDIAN LIFE: Customs and Traditions of 23 Tribes, Elsie Clews Parsons (ed.). 27 fictionalized essays by noted anthropologists examine religion, customs, government, additional facets of life among the Winnebago, Crow, Zuni, Eskimo, other tribes. 480pp. 6⅛ x 9¼. 27377-6

FRANK LLOYD WRIGHT'S DANA HOUSE, Donald Hoffmann. Pictorial essay of residential masterpiece with over 160 interior and exterior photos, plans, elevations, sketches and studies. 128pp. 9¼ x 10¾. 29120-0

THE MALE AND FEMALE FIGURE IN MOTION: 60 Classic Photographic Sequences, Eadweard Muybridge. 60 true-action photographs of men and women walking, running, climbing, bending, turning, etc., reproduced from rare 19th-century masterpiece. vi + 121pp. 9 x 12. 24745-7

1001 QUESTIONS ANSWERED ABOUT THE SEASHORE, N. J. Berrill and Jacquelyn Berrill. Queries answered about dolphins, sea snails, sponges, starfish, fishes, shore birds, many others. Covers appearance, breeding, growth, feeding, much more. 305pp. 5¼ x 8¼. 23366-9

ATTRACTING BIRDS TO YOUR YARD, William J. Weber. Easy-to-follow guide offers advice on how to attract the greatest diversity of birds: birdhouses, feeders, water and waterers, much more. 96pp. 5 3/16 x 8¼. 28927-3

MEDICINAL AND OTHER USES OF NORTH AMERICAN PLANTS: A Historical Survey with Special Reference to the Eastern Indian Tribes, Charlotte Erichsen-Brown. Chronological historical citations document 500 years of usage of plants, trees, shrubs native to eastern Canada, northeastern U.S. Also complete identifying information. 343 illustrations. 544pp. 6½ x 9¼. 25951-X

STORYBOOK MAZES, Dave Phillips. 23 stories and mazes on two-page spreads: Wizard of Oz, Treasure Island, Robin Hood, etc. Solutions. 64pp. 8¼ x 11. 23628-5

AMERICAN NEGRO SONGS: 230 Folk Songs and Spirituals, Religious and Secular, John W. Work. This authoritative study traces the African influences of songs sung and played by black Americans at work, in church, and as entertainment. The author discusses the lyric significance of such songs as "Swing Low, Sweet Chariot," "John Henry," and others and offers the words and music for 230 songs. Bibliography. Index of Song Titles. 272pp. 6½ x 9¼. 40271-1

MOVIE-STAR PORTRAITS OF THE FORTIES, John Kobal (ed.). 163 glamor, studio photos of 106 stars of the 1940s: Rita Hayworth, Ava Gardner, Marlon Brando, Clark Gable, many more. 176pp. 8⅜ x 11¼. 23546-7

BENCHLEY LOST AND FOUND, Robert Benchley. Finest humor from early 30s, about pet peeves, child psychologists, post office and others. Mostly unavailable elsewhere. 73 illustrations by Peter Arno and others. 183pp. 5⅜ x 8½. 22410-4

YEKL and THE IMPORTED BRIDEGROOM AND OTHER STORIES OF YIDDISH NEW YORK, Abraham Cahan. Film Hester Street based on *Yekl* (1896). Novel, other stories among first about Jewish immigrants on N.Y.'s East Side. 240pp. 5⅜ x 8½. 22427-9

SELECTED POEMS, Walt Whitman. Generous sampling from *Leaves of Grass*. Twenty-four poems include "I Hear America Singing," "Song of the Open Road," "I Sing the Body Electric," "When Lilacs Last in the Dooryard Bloom'd," "O Captain! My Captain!"–all reprinted from an authoritative edition. Lists of titles and first lines. 128pp. 5 3/16 x 8¼. 26878-0

THE BEST TALES OF HOFFMANN, E. T. A. Hoffmann. 10 of Hoffmann's most important stories: "Nutcracker and the King of Mice," "The Golden Flowerpot," etc. 458pp. 5⅜ x 8½. 21793-0

FROM FETISH TO GOD IN ANCIENT EGYPT, E. A. Wallis Budge. Rich detailed survey of Egyptian conception of "God" and gods, magic, cult of animals, Osiris, more. Also, superb English translations of hymns and legends. 240 illustrations. 545pp. 5⅜ x 8½. 25803-3

FRENCH STORIES/CONTES FRANÇAIS: A Dual-Language Book, Wallace Fowlie. Ten stories by French masters, Voltaire to Camus: "Micromegas" by Voltaire; "The Atheist's Mass" by Balzac; "Minuet" by de Maupassant; "The Guest" by Camus, six more. Excellent English translations on facing pages. Also French-English vocabulary list, exercises, more. 352pp. 5⅜ x 8½. 26443-2

CHICAGO AT THE TURN OF THE CENTURY IN PHOTOGRAPHS: 122 Historic Views from the Collections of the Chicago Historical Society, Larry A. Viskochil. Rare large-format prints offer detailed views of City Hall, State Street, the Loop, Hull House, Union Station, many other landmarks, circa 1904-1913. Introduction. Captions. Maps. 144pp. 9⅜ x 12¼. 24656-6

OLD BROOKLYN IN EARLY PHOTOGRAPHS, 1865-1929, William Lee Younger. Luna Park, Gravesend race track, construction of Grand Army Plaza, moving of Hotel Brighton, etc. 157 previously unpublished photographs. 165pp. 8⅞ x 11¾. 23587-4

THE MYTHS OF THE NORTH AMERICAN INDIANS, Lewis Spence. Rich anthology of the myths and legends of the Algonquins, Iroquois, Pawnees and Sioux, prefaced by an extensive historical and ethnological commentary. 36 illustrations. 480pp. 5⅜ x 8½. 25967-6

AN ENCYCLOPEDIA OF BATTLES: Accounts of Over 1,560 Battles from 1479 B.C. to the Present, David Eggenberger. Essential details of every major battle in recorded history from the first battle of Megiddo in 1479 B.C. to Grenada in 1984. List of Battle Maps. New Appendix covering the years 1967-1984. Index. 99 illustrations. 544pp. 6½ x 9¼. 24913-1

SAILING ALONE AROUND THE WORLD, Captain Joshua Slocum. First man to sail around the world, alone, in small boat. One of great feats of seamanship told in delightful manner. 67 illustrations. 294pp. 5⅜ x 8½. 20326-3

ANARCHISM AND OTHER ESSAYS, Emma Goldman. Powerful, penetrating, prophetic essays on direct action, role of minorities, prison reform, puritan hypocrisy, violence, etc. 271pp. 5⅜ x 8½. 22484-8

MYTHS OF THE HINDUS AND BUDDHISTS, Ananda K. Coomaraswamy and Sister Nivedita. Great stories of the epics; deeds of Krishna, Shiva, taken from puranas, Vedas, folk tales; etc. 32 illustrations. 400pp. 5⅜ x 8½. 21759-0

THE TRAUMA OF BIRTH, Otto Rank. Rank's controversial thesis that anxiety neurosis is caused by profound psychological trauma which occurs at birth. 256pp. 5⅜ x 8½. 27974-X

A THEOLOGICO-POLITICAL TREATISE, Benedict Spinoza. Also contains unfinished Political Treatise. Great classic on religious liberty, theory of government on common consent. R. Elwes translation. Total of 421pp. 5⅜ x 8½. 20249-6

MY BONDAGE AND MY FREEDOM, Frederick Douglass. Born a slave, Douglass became outspoken force in antislavery movement. The best of Douglass' autobiographies. Graphic description of slave life. 464pp. 5⅜ x 8½. 22457-0

FOLLOWING THE EQUATOR: A Journey Around the World, Mark Twain. Fascinating humorous account of 1897 voyage to Hawaii, Australia, India, New Zealand, etc. Ironic, bemused reports on peoples, customs, climate, flora and fauna, politics, much more. 197 illustrations. 720pp. 5⅜ x 8½. 26113-1

THE PEOPLE CALLED SHAKERS, Edward D. Andrews. Definitive study of Shakers: origins, beliefs, practices, dances, social organization, furniture and crafts, etc. 33 illustrations. 351pp. 5⅜ x 8½. 21081-2

THE MYTHS OF GREECE AND ROME, H. A. Guerber. A classic of mythology, generously illustrated, long prized for its simple, graphic, accurate retelling of the principal myths of Greece and Rome, and for its commentary on their origins and significance. With 64 illustrations by Michelangelo, Raphael, Titian, Rubens, Canova, Bernini and others. 480pp. 5⅜ x 8½. 27584-1

PSYCHOLOGY OF MUSIC, Carl E. Seashore. Classic work discusses music as a medium from psychological viewpoint. Clear treatment of physical acoustics, auditory apparatus, sound perception, development of musical skills, nature of musical feeling, host of other topics. 88 figures. 408pp. 5⅜ x 8½. 21851-1

THE PHILOSOPHY OF HISTORY, Georg W. Hegel. Great classic of Western thought develops concept that history is not chance but rational process, the evolution of freedom. 457pp. 5⅜ x 8½. 20112-0

THE BOOK OF TEA, Kakuzo Okakura. Minor classic of the Orient: entertaining, charming explanation, interpretation of traditional Japanese culture in terms of tea ceremony. 94pp. 5⅜ x 8½. 20070-1

LIFE IN ANCIENT EGYPT, Adolf Erman. Fullest, most thorough, detailed older account with much not in more recent books, domestic life, religion, magic, medicine, commerce, much more. Many illustrations reproduce tomb paintings, carvings, hieroglyphs, etc. 597pp. 5⅜ x 8½. 22632-8

SUNDIALS, Their Theory and Construction, Albert Waugh. Far and away the best, most thorough coverage of ideas, mathematics concerned, types, construction, adjusting anywhere. Simple, nontechnical treatment allows even children to build several of these dials. Over 100 illustrations. 230pp. 5⅜ x 8½. 22947-5

THEORETICAL HYDRODYNAMICS, L. M. Milne-Thomson. Classic exposition of the mathematical theory of fluid motion, applicable to both hydrodynamics and aerodynamics. Over 600 exercises. 768pp. 6⅛ x 9¼. 68970-0

SONGS OF EXPERIENCE: Facsimile Reproduction with 26 Plates in Full Color, William Blake. 26 full-color plates from a rare 1826 edition. Includes "The Tyger," "London," "Holy Thursday," and other poems. Printed text of poems. 48pp. 5¼ x 7. 24636-1

OLD-TIME VIGNETTES IN FULL COLOR, Carol Belanger Grafton (ed.). Over 390 charming, often sentimental illustrations, selected from archives of Victorian graphics–pretty women posing, children playing, food, flowers, kittens and puppies, smiling cherubs, birds and butterflies, much more. All copyright-free. 48pp. 9¼ x 12¼. 27269-9

PERSPECTIVE FOR ARTISTS, Rex Vicat Cole. Depth, perspective of sky and sea, shadows, much more, not usually covered. 391 diagrams, 81 reproductions of drawings and paintings. 279pp. 5⅜ x 8½. 22487-2

DRAWING THE LIVING FIGURE, Joseph Sheppard. Innovative approach to artistic anatomy focuses on specifics of surface anatomy, rather than muscles and bones. Over 170 drawings of live models in front, back and side views, and in widely varying poses. Accompanying diagrams. 177 illustrations. Introduction. Index. 144pp. 8⅜ x11¼. 26723-7

GOTHIC AND OLD ENGLISH ALPHABETS: 100 Complete Fonts, Dan X. Solo. Add power, elegance to posters, signs, other graphics with 100 stunning copyright-free alphabets: Blackstone, Dolbey, Germania, 97 more–including many lower-case, numerals, punctuation marks. 104pp. 8⅛ x 11. 24695-7

HOW TO DO BEADWORK, Mary White. Fundamental book on craft from simple projects to five-bead chains and woven works. 106 illustrations. 142pp. 5⅜ x 8. 20697-1

THE BOOK OF WOOD CARVING, Charles Marshall Sayers. Finest book for beginners discusses fundamentals and offers 34 designs. "Absolutely first rate . . . well thought out and well executed."–E. J. Tangerman. 118pp. 7¾ x 10⅝. 23654-4

ILLUSTRATED CATALOG OF CIVIL WAR MILITARY GOODS: Union Army Weapons, Insignia, Uniform Accessories, and Other Equipment, Schuyler, Hartley, and Graham. Rare, profusely illustrated 1846 catalog includes Union Army uniform and dress regulations, arms and ammunition, coats, insignia, flags, swords, rifles, etc. 226 illustrations. 160pp. 9 x 12. 24939-5

WOMEN'S FASHIONS OF THE EARLY 1900s: An Unabridged Republication of "New York Fashions, 1909," National Cloak & Suit Co. Rare catalog of mail-order fashions documents women's and children's clothing styles shortly after the turn of the century. Captions offer full descriptions, prices. Invaluable resource for fashion, costume historians. Approximately 725 illustrations. 128pp. 8⅜ x 11¼. 27276-1

THE 1912 AND 1915 GUSTAV STICKLEY FURNITURE CATALOGS, Gustav Stickley. With over 200 detailed illustrations and descriptions, these two catalogs are essential reading and reference materials and identification guides for Stickley furniture. Captions cite materials, dimensions and prices. 112pp. 6½ x 9¼. 26676-1

EARLY AMERICAN LOCOMOTIVES, John H. White, Jr. Finest locomotive engravings from early 19th century: historical (1804–74), main-line (after 1870), special, foreign, etc. 147 plates. 142pp. 11⅜ x 8¼. 22772-3

THE TALL SHIPS OF TODAY IN PHOTOGRAPHS, Frank O. Braynard. Lavishly illustrated tribute to nearly 100 majestic contemporary sailing vessels: Amerigo Vespucci, Clearwater, Constitution, Eagle, Mayflower, Sea Cloud, Victory, many more. Authoritative captions provide statistics, background on each ship. 190 black-and-white photographs and illustrations. Introduction. 128pp. 8⅞ x 11¾. 27163-3

LITTLE BOOK OF EARLY AMERICAN CRAFTS AND TRADES, Peter Stockham (ed.). 1807 children's book explains crafts and trades: baker, hatter, cooper, potter, and many others. 23 copperplate illustrations. 140pp. 4⅝ x 6. 23336-7

VICTORIAN FASHIONS AND COSTUMES FROM HARPER'S BAZAR, 1867–1898, Stella Blum (ed.). Day costumes, evening wear, sports clothes, shoes, hats, other accessories in over 1,000 detailed engravings. 320pp. 9⅜ x 12¼. 22990-4

GUSTAV STICKLEY, THE CRAFTSMAN, Mary Ann Smith. Superb study surveys broad scope of Stickley's achievement, especially in architecture. Design philosophy, rise and fall of the Craftsman empire, descriptions and floor plans for many Craftsman houses, more. 86 black-and-white halftones. 31 line illustrations. Introduction 208pp. 6½ x 9¼. 27210-9

THE LONG ISLAND RAIL ROAD IN EARLY PHOTOGRAPHS, Ron Ziel. Over 220 rare photos, informative text document origin (1844) and development of rail service on Long Island. Vintage views of early trains, locomotives, stations, passengers, crews, much more. Captions. 8⅞ x 11¾. 26301-0

VOYAGE OF THE LIBERDADE, Joshua Slocum. Great 19th-century mariner's thrilling, first-hand account of the wreck of his ship off South America, the 35-foot boat he built from the wreckage, and its remarkable voyage home. 128pp. 5⅜ x 8½. 40022-0

TEN BOOKS ON ARCHITECTURE, Vitruvius. The most important book ever written on architecture. Early Roman aesthetics, technology, classical orders, site selection, all other aspects. Morgan translation. 331pp. 5⅜ x 8½. 20645-9

THE HUMAN FIGURE IN MOTION, Eadweard Muybridge. More than 4,500 stopped-action photos, in action series, showing undraped men, women, children jumping, lying down, throwing, sitting, wrestling, carrying, etc. 390pp. 7⅞ x 10⅝. 20204-6 Clothbd.

TREES OF THE EASTERN AND CENTRAL UNITED STATES AND CANADA, William M. Harlow. Best one-volume guide to 140 trees. Full descriptions, woodlore, range, etc. Over 600 illustrations. Handy size. 288pp. 4½ x 6⅜. 20395-6

SONGS OF WESTERN BIRDS, Dr. Donald J. Borror. Complete song and call repertoire of 60 western species, including flycatchers, juncoes, cactus wrens, many more–includes fully illustrated booklet. Cassette and manual 99913-0

GROWING AND USING HERBS AND SPICES, Milo Miloradovich. Versatile handbook provides all the information needed for cultivation and use of all the herbs and spices available in North America. 4 illustrations. Index. Glossary. 236pp. 5⅜ x 8½. 25058-X

BIG BOOK OF MAZES AND LABYRINTHS, Walter Shepherd. 50 mazes and labyrinths in all–classical, solid, ripple, and more–in one great volume. Perfect inexpensive puzzler for clever youngsters. Full solutions. 112pp. 8⅛ x 11. 22951-3

PIANO TUNING, J. Cree Fischer. Clearest, best book for beginner, amateur. Simple repairs, raising dropped notes, tuning by easy method of flattened fifths. No previous skills needed. 4 illustrations. 201pp. 5⅜ x 8½. 23267-0

HINTS TO SINGERS, Lillian Nordica. Selecting the right teacher, developing confidence, overcoming stage fright, and many other important skills receive thoughtful discussion in this indispensible guide, written by a world-famous diva of four decades' experience. 96pp. 5⅜ x 8½. 40094-8

THE COMPLETE NONSENSE OF EDWARD LEAR, Edward Lear. All nonsense limericks, zany alphabets, Owl and Pussycat, songs, nonsense botany, etc., illustrated by Lear. Total of 320pp. 5⅜ x 8½. (Available in U.S. only.) 20167-8

VICTORIAN PARLOUR POETRY: An Annotated Anthology, Michael R. Turner. 117 gems by Longfellow, Tennyson, Browning, many lesser-known poets. "The Village Blacksmith," "Curfew Must Not Ring Tonight," "Only a Baby Small," dozens more, often difficult to find elsewhere. Index of poets, titles, first lines. xxiii + 325pp. 5⅜ x 8¼. 27044-0

DUBLINERS, James Joyce. Fifteen stories offer vivid, tightly focused observations of the lives of Dublin's poorer classes. At least one, "The Dead," is considered a masterpiece. Reprinted complete and unabridged from standard edition. 160pp. 5$\frac{3}{16}$ x 8¼. 26870-5

GREAT WEIRD TALES: 14 Stories by Lovecraft, Blackwood, Machen and Others, S. T. Joshi (ed.). 14 spellbinding tales, including "The Sin Eater," by Fiona McLeod, "The Eye Above the Mantel," by Frank Belknap Long, as well as renowned works by R. H. Barlow, Lord Dunsany, Arthur Machen, W. C. Morrow and eight other masters of the genre. 256pp. 5⅜ x 8½. (Available in U.S. only.) 40436-6

THE BOOK OF THE SACRED MAGIC OF ABRAMELIN THE MAGE, translated by S. MacGregor Mathers. Medieval manuscript of ceremonial magic. Basic document in Aleister Crowley, Golden Dawn groups. 268pp. 5⅜ x 8½. 23211-5

NEW RUSSIAN-ENGLISH AND ENGLISH-RUSSIAN DICTIONARY, M. A. O'Brien. This is a remarkably handy Russian dictionary, containing a surprising amount of information, including over 70,000 entries. 366pp. 4½ x 6⅛. 20208-9

HISTORIC HOMES OF THE AMERICAN PRESIDENTS, Second, Revised Edition, Irvin Haas. A traveler's guide to American Presidential homes, most open to the public, depicting and describing homes occupied by every American President from George Washington to George Bush. With visiting hours, admission charges, travel routes. 175 photographs. Index. 160pp. 8¼ x 11. 26751-2

NEW YORK IN THE FORTIES, Andreas Feininger. 162 brilliant photographs by the well-known photographer, formerly with *Life* magazine. Commuters, shoppers, Times Square at night, much else from city at its peak. Captions by John von Hartz. 181pp. 9¼ x 10¾. 23585-8

INDIAN SIGN LANGUAGE, William Tomkins. Over 525 signs developed by Sioux and other tribes. Written instructions and diagrams. Also 290 pictographs. 111pp. 6⅛ x 9¼. 22029-X

ANATOMY: A Complete Guide for Artists, Joseph Sheppard. A master of figure drawing shows artists how to render human anatomy convincingly. Over 460 illustrations. 224pp. 8⅜ x 11¼. 27279-6

MEDIEVAL CALLIGRAPHY: Its History and Technique, Marc Drogin. Spirited history, comprehensive instruction manual covers 13 styles (ca. 4th century through 15th). Excellent photographs; directions for duplicating medieval techniques with modern tools. 224pp. 8⅜ x 11¼. 26142-5

DRIED FLOWERS: How to Prepare Them, Sarah Whitlock and Martha Rankin. Complete instructions on how to use silica gel, meal and borax, perlite aggregate, sand and borax, glycerine and water to create attractive permanent flower arrangements. 12 illustrations. 32pp. 5⅜ x 8½. 21802-3

EASY-TO-MAKE BIRD FEEDERS FOR WOODWORKERS, Scott D. Campbell. Detailed, simple-to-use guide for designing, constructing, caring for and using feeders. Text, illustrations for 12 classic and contemporary designs. 96pp. 5⅜ x 8½. 25847-5

SCOTTISH WONDER TALES FROM MYTH AND LEGEND, Donald A. Mackenzie. 16 lively tales tell of giants rumbling down mountainsides, of a magic wand that turns stone pillars into warriors, of gods and goddesses, evil hags, powerful forces and more. 240pp. 5⅜ x 8½. 29677-6

THE HISTORY OF UNDERCLOTHES, C. Willett Cunnington and Phyllis Cunnington. Fascinating, well-documented survey covering six centuries of English undergarments, enhanced with over 100 illustrations: 12th-century laced-up bodice, footed long drawers (1795), 19th-century bustles, 19th-century corsets for men, Victorian "bust improvers," much more. 272pp. 5⅜ x 8¼. 27124-2

ARTS AND CRAFTS FURNITURE: The Complete Brooks Catalog of 1912, Brooks Manufacturing Co. Photos and detailed descriptions of more than 150 now very collectible furniture designs from the Arts and Crafts movement depict davenports, settees, buffets, desks, tables, chairs, bedsteads, dressers and more, all built of solid, quarter-sawed oak. Invaluable for students and enthusiasts of antiques, Americana and the decorative arts. 80pp. 6½ x 9¼. 27471-3

WILBUR AND ORVILLE: A Biography of the Wright Brothers, Fred Howard. Definitive, crisply written study tells the full story of the brothers' lives and work. A vividly written biography, unparalleled in scope and color, that also captures the spirit of an extraordinary era. 560pp. 6⅛ x 9¼. 40297-5

THE ARTS OF THE SAILOR: Knotting, Splicing and Ropework, Hervey Garrett Smith. Indispensable shipboard reference covers tools, basic knots and useful hitches; handsewing and canvas work, more. Over 100 illustrations. Delightful reading for sea lovers. 256pp. 5⅜ x 8½. 26440-8

FRANK LLOYD WRIGHT'S FALLINGWATER: The House and Its History, Second, Revised Edition, Donald Hoffmann. A total revision–both in text and illustrations–of the standard document on Fallingwater, the boldest, most personal architectural statement of Wright's mature years, updated with valuable new material from the recently opened Frank Lloyd Wright Archives. "Fascinating"–*The New York Times*. 116 illustrations. 128pp. 9¼ x 10¾. 27430-6

PHOTOGRAPHIC SKETCHBOOK OF THE CIVIL WAR, Alexander Gardner. 100 photos taken on field during the Civil War. Famous shots of Manassas Harper's Ferry, Lincoln, Richmond, slave pens, etc. 244pp. 10⅝ x 8¼. 22731-6

FIVE ACRES AND INDEPENDENCE, Maurice G. Kains. Great back-to-the-land classic explains basics of self-sufficient farming. The one book to get. 95 illustrations. 397pp. 5⅜ x 8½. 20974-1

SONGS OF EASTERN BIRDS, Dr. Donald J. Borror. Songs and calls of 60 species most common to eastern U.S.: warblers, woodpeckers, flycatchers, thrushes, larks, many more in high-quality recording. Cassette and manual 99912-2

A MODERN HERBAL, Margaret Grieve. Much the fullest, most exact, most useful compilation of herbal material. Gigantic alphabetical encyclopedia, from aconite to zedoary, gives botanical information, medical properties, folklore, economic uses, much else. Indispensable to serious reader. 161 illustrations. 888pp. 6½ x 9¼. 2-vol. set. (Available in U.S. only.) Vol. I: 22798-7
Vol. II: 22799-5

HIDDEN TREASURE MAZE BOOK, Dave Phillips. Solve 34 challenging mazes accompanied by heroic tales of adventure. Evil dragons, people-eating plants, blood-thirsty giants, many more dangerous adversaries lurk at every twist and turn. 34 mazes, stories, solutions. 48pp. 8¼ x 11. 24566-7

LETTERS OF W. A. MOZART, Wolfgang A. Mozart. Remarkable letters show bawdy wit, humor, imagination, musical insights, contemporary musical world; includes some letters from Leopold Mozart. 276pp. 5⅜ x 8½. 22859-2

BASIC PRINCIPLES OF CLASSICAL BALLET, Agrippina Vaganova. Great Russian theoretician, teacher explains methods for teaching classical ballet. 118 illustrations. 175pp. 5⅜ x 8½. 22036-2

THE JUMPING FROG, Mark Twain. Revenge edition. The original story of The Celebrated Jumping Frog of Calaveras County, a hapless French translation, and Twain's hilarious "retranslation" from the French. 12 illustrations. 66pp. 5⅜ x 8½. 22686-7

BEST REMEMBERED POEMS, Martin Gardner (ed.). The 126 poems in this superb collection of 19th- and 20th-century British and American verse range from Shelley's "To a Skylark" to the impassioned "Renascence" of Edna St. Vincent Millay and to Edward Lear's whimsical "The Owl and the Pussycat." 224pp. 5⅜ x 8½. 27165-X

COMPLETE SONNETS, William Shakespeare. Over 150 exquisite poems deal with love, friendship, the tyranny of time, beauty's evanescence, death and other themes in language of remarkable power, precision and beauty. Glossary of archaic terms. 80pp. 5 3/16 x 8¼. 26686-9

THE BATTLES THAT CHANGED HISTORY, Fletcher Pratt. Eminent historian profiles 16 crucial conflicts, ancient to modern, that changed the course of civilization. 352pp. 5⅜ x 8½. 41129-X